The Experimental Basis of Physical Chemistry

The Experimental Basis of Physical Chemistry

G. Van Praagh

B.Sc (London)   Ph.D (Cantab.)   F.R.I.C.
Formerly Head of the Science Department, Christ's Hospital

John Murray   London

Printed in Great Britain by Butler & Tanner Ltd, Frome and London

7195 1846 6

Preface

This book is based on the author's earlier *Physical Chemistry, Experimental and Theoretical* (Cambridge University Press), which has been much revised and to which several new topics have been added. It is hoped that the book will meet the need for a laboratory manual of experiments in physical chemistry suitable for school and college use. The limited time and equipment usually available render impracticable many experiments in physical chemistry included in more advanced courses. A number of these have here been simplified and adapted. Most of these have been tried out by students and in many cases specimen results are quoted that have been obtained using the methods and apparatus described in the text.

Many students do not begin to get a real understanding of a new subject until they observe the phenomena for themselves in the laboratory. In this book, therefore, each topic is introduced by means of a few short experiments designed to demonstrate in a simple manner the subject-matter of the section. These experiments are followed by brief theoretical sections. An outline of the historical development of the subject is sometimes included, because this often throws considerable light on the present state of knowledge and theory. At the end of each section, a number of other experiments are described, some of which at least the student should perform for himself. More experiments are included than would normally be done by any one student in an introductory course.

The choice of topics to be included in the subject of physical chemistry is somewhat arbitrary: the selection made here is governed by the suitability of the various topics for experimental treatment in school or college laboratory. The mathematics has been kept simple: emphasis is rather laid on the physico-chemical phenomena themselves and their significance in various branches of applied science. There are already many written exercises and questions published in other textbooks and therefore none have been included here. Chemicals are named according to the I.U.P.A.C. convention, with the exception of a few whose familiar names are more generally used. Although most units are S.I., some in current use have been retained where more appropriate. A table for the conversion of calories to joules is appended by permission of the Nuffield Physical Science Project.

The author of a textbook derives many ideas from others and it is difficult to give all the credits where they are due. I owe a great deal to colleagues in the teaching profession with whom I have worked and discussed matters during many years. I am indebted to J. Bradley for several ideas that also appeared in the original version of this book, in particular for his exposition of Cannizzaro's proof of the formula for hydrogen gas and for Helmholtz's views on the stability of ions (see J. Bradley, *S.S.R.* 1942, **91**, 23; 1957, **137**, 34 and 1958, **138**, 34). More recently I have learnt much from colleagues in the Nuffield Science Teaching Project. Discussions with them have led to the modification of some sections and to the inclusion of some new experiments that appear in the Nuffield publications (Expts. 1d-1 to 6, 9c-8 and 11a-3). For permission to use these, I am indebted to the Nuffield Foundation. Fig. 6 is reproduced by permission of C. V. Platts; Plate 5a is taken from L. J. Griffin, *Philosophical Magazine*, Ser. 7, vol. xli (Taylor & Francis), and Plate 5b from A. R. Verma, *Crystal Growth and Dislocations* (Butterworths). The following figures were adapted from diagrams in various publications and acknowledgement is due as shown: Figs. 26, 28, 29, A. F. Wells, *Structural Inorganic Chemistry* (O.U.P.); Figs. 24, 30, 31, C. W. Bunn, *Crystals* (Academic Press); Fig. 77, the Permutit Co. Ltd; Fig. 40, Unilever Educational Booklets; Figs. 32, 33, the Faraday Society. I am also indebted to T. A. H. Peacocke for advice on Chapter 1(d).

I also wish to express my thanks to Dr. D. J. Waddington for reading the typescript and making many valuable suggestions. My former pupil J. F. Duke drew many of the diagrams. I should also like to thank the publishers for the trouble they have taken in preparing the material for the press.

G. VAN PRAAGH
June 1969

Contents

Plates

5 Spiral Growth Sheets

(a) Prism surface of beryl (Griffin) ($\times 700$)
(b) Spiral on a SiC crystal (Verma) (130)

Facing page 96

6 Solid solutions

(a) 50% copper-nickel alloy, etched to show dendritic structure ($\times 86$)
(b) Silver-copper alloy showing dendrites of copper (containing a little silver in solid solution) lying in a silver-copper eutectic ($\times 200$)
(c) Zoned plagioclase felspar (solid solutions of sodium and calcium alumino-silicates) (crossed nicols, $\times 50$)

Facing page 97

7 Leisegang rings

(a) Magnesium hydroxide bands in gelatine ($\times 4/5$)
(b) Silver chromate bands in gelatine ($\times 3/2$)
(c) Banded agate ($\times 1$)

Experiments

Chapter 1

Chapter 2

Chapter 3

1 Atoms and molecules

(a) History of the atomic theory

Expt. 1a-1

Put 80–90 cm³ of distilled water in a measuring cylinder, read the volume, and pour in about a teaspoonful of anhydrous sodium sulphate. Without stirring, quickly read the level of water in the cylinder. Shake up the mixture to dissolve the salt and when solution is complete, again read the volume. The result of this and similar experiments was one of the observations that Dalton used in developing his atomic theory.

The view that matter is not continuous but consists of minute indivisible particles was probably first held by the Hindus as early as 1200 B.C. It was put forward by the Greek philosopher Leukippus in about 500 B.C. and developed by Democritus about 60 years later. Just over half a century B.C., views which sound very similar to those held today were worked out by Lucretius in a long Latin poem entitled *De rerum natura*. He regarded the ultimate particles of matter as indestructible and always in motion. In a solid they are packed closely together, in a liquid they are loosely packed, and in a gas they are widely separated and free to move about. The physical properties of a substance depended upon the manner in which its constituent atoms are arranged.

In the seventeenth century it was the custom to think in terms of atoms and to use them to explain observed phenomena. Thus Sir Isaac Newton, about 1680, wrote:

'It seems to me probable that God in the first beginning formed matter in solid, massy, hard, impenetrable, moveable particles, and that these primitive solids are incomparably harder than the porous bodies compounded of them; even so very hard as never to wear or break in pieces, no ordinary power being able to divide what God Himself made one in the first creation.'

Newton was able to show that Boyle's law follows from the hypothesis that gases consist of atoms repelling one another with a force inversely proportional to their distances apart. A notable application of the atomic hypothesis to chemistry was made by the two Irish chemists, Bryan and William Higgins, about 1790. They attempted to find the number of atoms of reacting substances which combined to form single atoms of the products. Adopting Newton's view that the atoms repelled each other, they assumed that the *minimum* number of atoms would unite, usually two. Combination in multiple proportions was recognized, but it was presumed that combination of one atom of each component led to the compound of greatest stability.

About 1801-2 John Dalton formulated his Chemical Atomic Theory. It found wide acceptance at the time, and, although further knowledge has modified and extended Dalton's conceptions, it formed the basis of theoretical chemistry for over a century. The theory undoubtedly arose through the influence of Newton's ideas and was developed from a consideration of the physical properties of gases, liquids and solutions. Dalton's contemporaries held that the theory was put forward to explain the law of multiple proportions, but it is clear from Dalton's notebooks, discovered in 1895, that this was not so, and that the law of multiple proportions was deduced from his atomic hypothesis. Dalton's postulates may be summarized as follows:

(1) The atoms of an element are indestructible, all alike in size and mass, and differ from those of any other element.

(2) Compound atoms result from the combination of simple atoms in simple numerical proportions.

The following extract from his writings is an example of the reasoning that led Dalton to these conclusions: 'It is scarcely possible to conceive how the aggregates of dissimilar particles could be so uniformly alike. If some particles of water were heavier than others, if a parcel of liquid was constituted principally of these heavy particles, it must be supposed to affect the specific gravity of the mass, a circumstance not known. We must conclude that every particle of water is like every other particle of water.' He then applied a similar reasoning to the particles of oxygen and hydrogen, of which each particle of water consists.

Of chemical change he says: 'Analysis and synthesis go no further than the separation of particles and their reunion.'

Dalton represented his atoms by symbols, e.g. ○ for an atom of oxygen, ⊙ for an atom of hydrogen, ● for an atom of carbon, and so on. It was to the use of these symbols, which made the ideas of chemical combination vivid and easy to grasp, that the success of the theory was in some part due. Berzelius later used the first letter of the name of the element in a circle, instead of Dalton's symbols, e.g. Ⓒ, Ⓗ, Ⓞ, etc., but he soon omitted the circles, leaving symbols similar to those at present in use.

THE LAWS OF CHEMICAL COMBINATION

Dalton's atomic theory provided an explanation for the laws of chemical combination, which were being developed from experimental work at the close of the eighteenth century and the beginning of the nineteenth.

The law of conservation of mass

In 1630 Jean Rey, in his essay entitled 'Heaviness is so closely united to the primary matter of the elements, that when these are changed one into the other they always retain the same weight', wrote: 'It is reason which leads me to give a flat denial to the erroneous maxim that the elements mutually undergoing change, lose or gain weight. With the arms of reason I boldly enter the lists to combat this error, and to sustain that weight is so closely united to the primary matter of the elements that they can never be deprived of it. The weight with which each portion of matter was endowed at the cradle, will be carried by it to the grave!'

Lavoisier stated the law in 1789 in the words: 'In every operation there is an equal quantity of matter before and after the operation.' The law was later tested by Landolt in a series of experiments from 1893 to 1908, and found to hold within the limits of his experimental error, viz. one part in ten million.

In radioactive transformations the mass of the products is seldom equal to that of the reactants, some mass having been transformed into an equivalent amount of energy.

The law of constant composition or fixed proportions

This law, tacitly assumed by Black and Lavoisier, was stated and tested by Proust (Professor of Chemistry at Madrid) in 1799–1802: 'The elements combine together in fixed proportions by weight.'

The law was challenged by Berthollet, who, in 1803 under Napoleon, took the place of Lavoisier as the recognized leader of French science. He maintained that the combining proportions could be varied by the chemist, and that fixed proportions only result when crystallization from mixtures occurs, which is the method so frequently used in the preparation of chemical compounds. He also instanced the gradual change in the colour and weight of lead and iron on fusion in air as progressive oxidation occurs. The confusion arose largely through lack of a clear distinction between a *compound* and a *solution*. In 1804–8 Proust used the law to make this distinction: 'The attraction that makes sugar dissolve in water may or may not be the same as that which makes a definite quantity of carbon and hydrogen "dissolve" in another definite quantity of oxygen to make sugar, but the two sorts of attraction are so different in their results that it is impossible to confuse them' (1806).

The accuracy of the law was tested by Stas (1865) and shown to hold up to the limit of his experimental accuracy, namely 0.002%. The law is now known not to hold for many so-called non-stoichiometric substances, e.g. the composition of FeO varies from $FeO_{1.055}$ to $FeO_{1.19}$.

The law of multiple proportions

As stated previously, this law was deduced by Dalton about 1803 from his atomic theory. It was first stated and proved experimentally for the oxides of lead, copper, sulphur and iron, in 1810 by Berzelius, Professor of Medicine and Pharmacy at Stockholm and the most distinguished chemical philosopher and analyst of his time: 'When two elements combine to form more than one compound, the weights of one which combine with identical weights of the other, are in simple numerical proportion.'

Thus it is clear on the atomic theory that if in 'carbonic oxide', © ©, the proportion of carbon to oxygen is 3 : 4, then in 'carbonic anhydride', © © ©, it will be 3 : 8. The accuracy of the law was tested by Stas in 1849 with the oxides of carbon and found to hold within 0.015%. Consideration of the carbon/hydrogen ratio in organic compounds shows that Berzelius' use of the phrase 'simple numerical proportion' can no longer be justified.

EQUIVALENT WEIGHTS

The conception of 'equivalent' or 'combining weights' was recognized many years earlier. Cavendish knew that the quantities of nitric acid and sulphuric acid which neutralized equal weights of potash, would also be neutralized by equal weights of marble. He described the two quantities of acids in 1766 as *equivalent* to one another. In 1819 Berzelius drew up a table of equivalent weights on the basis of Richter's work.

The *accurate* determination of several equivalent weights (on which depended the values of atomic weights, as will be seen later) was carried out by Stas in 1860 by a method suggested by Berzelius. It involves four steps, and is suitable for determining the equivalents of silver, the halogens and the alkali metals:

(i) When 127·2125 g of potassium chlorate are heated, 77·4023 g of potassium chloride are left and 49·8102 g of oxygen come off. Since potassium chlorate contains six equivalents of oxygen, the equivalent weight of potassium chloride is

$$\frac{77{\cdot}4023 \times 48{\cdot}00}{49{\cdot}8102}, \quad \text{i.e. } 74{\cdot}59 \text{ g.}$$

(ii) When an equivalent of potassium chloride is completely precipitated with silver nitrate as silver chloride, 143·397 g of the latter are obtained.

(iii) When pure silver is heated in chlorine, 143·397 g of silver chloride are found to be formed from 107·943 g of silver, and thus contain 35·454 g of chlorine. These are, therefore, the equivalent weights of silver and chlorine respectively.

(iv) The equivalent weight of potassium is therefore 74·59 − 35·454 g, i.e. 39·14 g.

The work was repeated in 1890 by T. W. Richards of Harvard University, using smaller quantities of material, which could be more easily purified than larger quantities. He employed silica vessels and electrical methods of heating. In this way he obtained an accuracy of 1 part in 100,000.

ILLUSTRATIONS OF THE LAW OF MULTIPLE PROPORTIONS
Expt. 1a-2 The two chlorides of copper

Weigh accurately two pieces of pure copper foil (about 1 and 2 g). Dissolve the smaller piece in the least quantity of concentrated nitric acid in a small round-bottomed flask. Evaporate to dryness, taking care to avoid loss by spray, and heat the nitrate gently until it is all converted to black copper oxide. Dissolve this by adding 30–40 cm³ of concentrated hydrochloric acid. Fit a Bunsen valve to the flask, add the larger piece of copper and boil in the fume cupboard until the solution is a pale amber colour (about 30 min). Remove the piece of copper, wash with distilled water and alcohol and weigh when dry. The loss in weight should be found to be equal to the weight of the smaller piece of copper used originally.

The copper(II) chloride formed from the black copper oxide is converted by the second piece of copper into copper(I) chloride, which thus contains the same weight of chlorine but twice the weight of copper as that present in the copper(II) chloride.

Expt. 1a-3   The two chlorides of mercury

Weigh two small beakers of about 150 cm³ capacity. Put about 5 g of mercury(I) chloride into one and about 5 g of mercury(II) chloride into the other, and weigh again. Reduce the two chlorides to mercury by warming them with hypophosphorous acid on a water-bath. (Use about 30 cm³ of water and 15 cm³ of acid.) Pour off the acid and wash the globules of mercury with distilled water. Remove most of the residual water with filter paper and finally dry the mercury by warming the beakers on a water-bath. Weigh them and hence calculate the weights of chlorine combined with equal weights of mercury in the two chlorides.

(b) The relative weights of the atoms

For many years equivalent weights were taken as proportional to atomic weights, and this practice led to a confusion which impeded the progress of theoretical chemistry for about half a century. If atoms only combined according to the simplest of Dalton's suggestions, namely 1 atom of A with 1 of B, then it is clear that the equivalent weight is a measure of the atomic weight. But it was known from the existence of more than one compound between the same two elements that this simple rule was inadequate, and that there also occur combinations of the type, 1 atom of A with 2 of B, and so on. Without knowing the ratio of the number of atoms of A to those of B in a given compound, it was impossible to determine the relative atomic weights of A and B. For example, gravimetric analysis of water showed that 1 g of hydrogen combines with 7·94 g of oxygen. Now if the 'compound atom' of water is Ⓗ Ⓞ, the atomic weight of oxygen, relative to the weight of an atom of hydrogen taken as unity, is 7·94. But if the 'compound atom' of water is Ⓗ Ⓞ Ⓗ, the oxygen atoms in the sample of water analysed are 7·94 times as heavy as the hydrogen atoms, of which there are twice as many. Thus, 1 atom of oxygen is 7·94 × 2 = 15·88 times as heavy as an atom of hydrogen.

Hence some additional information was necessary before atomic weights could be fixed with certainty—information about the relative numbers in which atoms are mutually combined in 'compound atoms'. Gay Lussac in 1809 and Avogadro in 1811 provided the 'missing links', but their significance was not fully realized until the time of Cannizzaro (1858).

Gay Lussac's law of combining volumes

In 1781 Cavendish measured the volumes in which oxygen and hydrogen combine to form water, and found that 1 vol. of oxygen combined with 2 vol. of hydrogen. In 1805 Gay Lussac repeated this determination and was so struck by the simplicity of the relationship, in such contrast with the complex ratio of the combining weights, that, in 1809, he extended his observations to the combinations of other gases. He also made observations on the volumes of the gas produced in a reaction, and formulated the law: 'Gases combine in simple proportions by volume, and the volumes of the gaseous products are simply related to the volume of the reacting gases, at constant temperature.'

It was clear from Boyle's law (1725) and Gay Lussac's discovery (1802) that all gases 'are expanded equally by the same degrees of heat', that the volume of a gas depends upon physical factors (temperature and pressure) rather than on the chemical nature of its constituent particles. It was recognized (from such facts as 1 cm³ of water occupies about 1200 cm³ when converted into steam) that these particles are widely separated in space, and that their chemical nature was unlikely to have much relation to the volume occupied by the gas. It was such considerations that must have given Dalton what he refers to in 1808 as 'a confused idea that I had at one time that the same number of particles of all gases occupied the same volume'. However, he abandoned this idea, and concluded wrongly that 'no two elastic fluids, probably, have the same number of particles, either in the same volume or the same weight'.

When Gay Lussac put forward his law of combining volumes in the following year, Dalton refused to recognize its validity, and in 1810 he wrote: 'The truth is, I believe, that gases do not unite in equal or exact measures in any one instance; when they appear to do so, it is owing to the inaccuracy of our experiments.'

AVOGADRO'S HYPOTHESIS

The hypothesis that equal volumes of different gases contain the same number of atoms, rejected by Dalton in 1808, was revived in a modified form by Amadeo Avogadro (Professor of Physics at Turin) in 1811. In order to account for the facts summarized in Gay Lussac's law, he assumed that equal volumes of different gases, when their temperatures and pressures are equal, contain the same number of freely moving particles; but Avogadro saw that these gas particles need not be, as Dalton assumed, single atoms, but might be little groups or clusters of atoms. Clerk Maxwell defined a 'molecule' of a gas as 'that small portion of matter which moves about as a whole so that its parts, if it has any, do not part company during the motion of agitation of the gas'.

Avogadro's hypothesis was postulated to bring together the two statements that (1) gases combine in simple proportions by weight (atomic theory), and (2) gases combine in simple proportions by volume (Gay Lussac's law). Dalton's attempted solution, that equal volumes of all gases contained the same number of atoms, is an unnecessary oversimplification and breaks down. Avogadro postulates a relation of extreme simplicity between number of particles and volumes, but each particle itself is a compound of a small number of atoms.

Avogadro published his ideas in the *Journal de Physique* for 1811, but it was not until 47 years later that their great importance to chemistry was indicated by Cannizzaro. In his hands, Avogadro's hypothesis came to be realized as the guiding principle that was needed to place theoretical chemistry on a sound foundation.

Application of Avogadro's hypothesis to gas reactions

Avogadro knew that
    2 cm^3 of hydrogen + 1 cm^3 of oxygen form 2 cm^3 of steam,
and concluded that
    2 mol. of hydrogen + 1 mol. of oxygen form 2 mol. of steam. (1)

From similar volume relationships he deduced that

1 mol. of oxygen + 1 mol. of nitrogen form 2 mol. of nitric oxide. (2)

2 mol. of ammonia are decomposed into 1 mol. of nitrogen and 3 mol. of hydrogen, (3)

1 mol. of chlorine + 1 mol. of hydrogen form 2 mol. of hydrogen chloride. (4)

The molecules of all the gaseous *elements* discussed by Avogadro were found to be divisible into two parts, but into no more than two. He therefore concluded that the indivisible 'atom' of these gases was probably the half-molecule. Further evidence of the truth of this view will be given later.

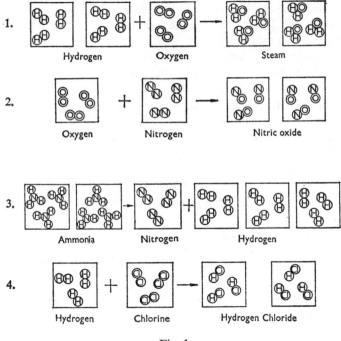

Fig. 1

The volume changes occurring in the above gas reactions may be represented by the diagrams in Fig. 1, in which equal volumes contain the same number of particles (viz. 3).

It follows at once from Avogadro's hypothesis that the ratio of the densities of two gases is equal to the ratio of the weights of their molecules. Hence the density of a gas relative to that of hydrogen as unity is equal to the weight of a molecule of the gas relative to that of hydrogen as unity. It is usual to express the weights of atoms and molecules as multiples of the weight of the lightest atom, namely, hydrogen. If the hydrogen molecule contains n atoms, it follows that the molecular weight of a gas is equal to its relative density multiplied by n. We have said above that n is probably 2, and further evidence for this follows.

Expt. 1b-1   The volume composition of ammonia
(Hoffmann, adapted by Fowles)

Collect a burette full of chlorine by displacement over strong brine. The chlorine is best prepared by dropping concentrated hydrogen chloride on to potassium permanganate. Insert a rubber stopper in the burette, and open the tap momentarily to reduce the pressure to atmospheric. Invert the burette, and place its tip under 0·880 ammonia in a mortar. Reduce the pressure in the burette by wetting the outside with a little strong ammonia. Open the tap cautiously and admit about 1 cm³ of ammonia. There is a small flash as the chlorine reacts to form nitrogen and white fumes of ammonium chloride. Cautiously admit a little more ammonia until the reaction is complete, but avoid excess. Neutralize the ammonia by letting in some half-concentrated hydrochloric acid. Adjust the pressure to atmospheric by opening the tap under water in a tall jar, cool the burette to room temperature, and take the reading. Finally, measure the volume between the zero mark and the bung and between the 50 cm³ mark and the tap, and calculate the volumes of the chlorine used and the nitrogen formed. Ammonia gas must then consist of this volume of nitrogen and the amount of hydrogen necessary to combine with the measured volume of chlorine, which we already know occupies a volume equal to the volume of the chlorine.

Expt. 1b-2   Determination of relative densities
(Victor Meyer's method)

The simplified apparatus shown in Fig. 2 may be used: A, a tin can; B, a tin lid cut in half; C, a long-necked flask; D, dry sand; E, rubber stopper.

The small stopper E should fit into a tube approximately 6 mm internal diameter. The small displacement of air on inserting the stopper is then negligible.

Make a small capsule about 12 mm long from glass tubing less than 6 mm external diameter, as illustrated. Weigh it empty, and then, by gently warming and cooling, draw in about 0·2–0·3 g of chloroform. Weigh again, and seal off at the capillary.

Keep the water in the can *A* boiling gently. When air bubbles have ceased coming from the delivery tube, place the latter in position under the burette. The length of tube below the water should be kept to a minimum. Then perform the following series of operations as rapidly as possible: break the neck of the chloroform tube,

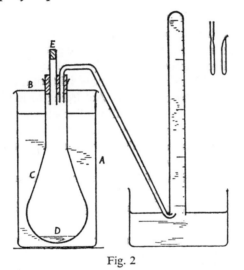

Fig. 2

remove the bung *E*, drop in the chloroform tube, replace the bung. The chloroform should vaporize rapidly and displace air which collects in the burette. When displacement of air ceases, transfer the burette to a tall jar of water, equalize the water levels inside and outside, and note the volume of air collected. Note the atmospheric temperature and pressure, and correct the measured volume to s.t.p. This is the volume that the vapour formed from the known weight of chloroform would occupy. Hence the density of chloroform vapour at s.t.p. may be calculated and compared with that of hydrogen. The relative density should be about 60.

THE DETERMINATION OF ATOMIC WEIGHTS

The scheme developed by Cannizzaro in his book *A Sketch of a Course of Chemical Philosophy* (1858) takes as the definition of atomic weight 'the smallest weight of the element found in a molecular weight of any of its compounds'. By measuring the relative densities of the vapours of a number of volatile compounds of the element, we have, according to Avogadro's hypothesis, a series of numbers proportional to the molecular weights of the vapours. We have seen that there was evidence for believing that the hydrogen molecule contains two atoms, and this leads to the relation

molecular weight $= 2 \times$ relative density.

However, Cannizzaro's method offers more convincing evidence for the diatomicity of the hydrogen molecule. Assuming that the hydrogen molecule contains n atoms, the molecular weight of any gas $= n \times$ its relative density. From gravimetric analyses of the compounds whose relative densities have been measured, the weights of the element present in each compound can be calculated. According to the law of equivalents, these weights will be in the ratios of small whole numbers. The least weight (or possibly the highest common factor) will be the atomic weight of the element. Cannizzaro's figures for compounds of hydrogen are given in the table below.

It is clear that the weight of hydrogen in a molecular weight of its compounds is either $\dfrac{n}{2}$, n or $\dfrac{3n}{2}$, indicating that $\dfrac{n}{2}$ is the weight of one

Compound	Density relative to hydrogen = 1	Mol. wt. if atomicity of hydrogen = n	Percentage of hydrogen	Weight of hydrogen in mol. wt. of compound
Hydrogen	1	n	100	n
Hydrogen chloride	18·25	18·25n	2·73	$\dfrac{2\cdot73 \times 18\cdot25n}{100} = \dfrac{n}{2}$
Hydrogen bromide	40·5	40·5n	1·23	$\dfrac{1\cdot23 \times 40\cdot5n}{100} = \dfrac{n}{2}$
Hydrogen iodide	64	64n	0·78	$\dfrac{0\cdot78 \times 64n}{100} = \dfrac{n}{2}$
Water vapour	9	9n	11·1	$\dfrac{11\cdot1 \times 9n}{100} = n$
Hydrogen sulphide	17	17n	5·88	$\dfrac{5\cdot88 \times 17n}{100} = n$
Ammonia	8·5	8·5n	17·6	$\dfrac{17\cdot6 \times 8\cdot5n}{100} = \dfrac{3n}{2}$
Arsine	39	39n	3·85	$\dfrac{3\cdot85 \times 39n}{100} = \dfrac{3n}{2}$
Phosphine	17	17n	8·86	$\dfrac{8\cdot86 \times 17n}{100} = \dfrac{3n}{2}$

atom, and that molecules containing n parts of hydrogen contain 2 atoms, etc. Now if the weight of an atom of hydrogen is taken as the unit for atomic and molecular weights, $\dfrac{n}{2} = 1$, and therefore n, the atomicity of the hydrogen molecule, is 2. It is worth pointing out that the actual value of the atomic weight of hydrogen might be anything, but the above proves that the molecule contains 2 atoms, whatever their weight.

The method can now be applied to other elements, taking the molecular weights of the compounds as twice their relative densities, and the following example will suffice to illustrate the method. *Cannizzaro* applied it to many elements, but the method is, of course, restricted to those elements that form a number of volatile compounds.

Compound	Relative density	Mol. wt.	Percentage of carbon in compound	Weight of carbon in a mol. wt. of compound
Carbon dioxide	22	44	27·27	$\dfrac{27 \cdot 27 \times 44}{100} = 12$
Carbon disulphide	38	76	15·8	$\dfrac{15 \cdot 8 \times 76}{100} = 12$
Ethyl alcohol	23	46	52·17	$\dfrac{52 \cdot 17 \times 46}{100} = 24$
Ethyl ether	37	74	64·86	$\dfrac{64 \cdot 86 \times 74}{100} = 48$
Chloroform	60	120	10·04	$\dfrac{10 \cdot 04 \times 120}{100} = 12$
Benzene	39	78	92·3	$\dfrac{92 \cdot 3 \times 78}{100} = 72$
Carbon tetrachloride	77	154	7·8	$\dfrac{7 \cdot 8 \times 154}{100} = 12$

It can be seen that the smallest weight of carbon in a gram-molecule of any of the compounds is 12; hence this is taken as the atomic weight of carbon. The numbers of carbon atoms in a molecule of the compounds are then respectively 1, 1, 2, 4, 1, 6, 1.

In thus perceiving the significance of Avogadro's hypothesis and in applying it as outlined above, Cannizzaro solved the problem expounded on p. 7 and placed the system of atomic and molecular weights on a sound basis.

Since it is not possible to measure relative densities with accuracy, owing to deviations from the gas laws, Cannizzaro's method establishes the approximate value of the atomic weight only, and the exact figure had to be obtained by multiplying the equivalent weight by the appropriate small whole number. For example, Cannizzaro's method gives 16 for the approximate atomic weight of oxygen. The equivalent weight, as measured by Dumas, and by Morley, is 7·935. Now $\frac{16}{7·935} \fallingdotseq 2$.

Hence the atomic weight relative to H = 1 is 7·935 × 2 = 15·87.

This 'small whole number' by which the equivalent weight must be multiplied in order to give the atomic weight is often called the 'valency' of the atom. The valency of an element, thus defined, is the number of atoms of hydrogen which one atom of the element will combine with or displace. In the above example, the valency of oxygen is 2, and the formula for the molecule of its compound with hydrogen is ⒣—Ⓞ—⒣. Thus from a knowledge of the atomic and equivalent weights, the valency can be obtained, but this is not necessarily equal to the number of bonds formed by the atom. The word 'valency' is often confusingly used in these two senses.

Other methods of atomic weight determination

(i) The metals form very few volatile compounds, so Cannizzaro's method is not applicable to the determination of their atomic weights. As early as 1819, Dulong and Petit, using the then accepted values for the atomic weights of the metals (many of which happened to be the right multiple of the equivalent weights) discovered that the product of the atomic weight and the specific heat lay within a small range, of which the mean value was 6·4. By means of a few elements (iodine and mercury) which form some volatile compounds and of which the specific heats can also be measured, it was shown that Dulong and Petit's rule held for these elements when their atomic weights obtained by Cannizzaro's method were used. Hence it can now be used with more confidence to provide approximate values of the atomic weights of metals from their specific heats. The exact values must again be obtained from the equivalent weights. For example:

The specific heat of copper   = 0·0936

The approximate atomic weight, from Dulong and Petit's rule,

$$= \frac{6·4}{0·0936} = 68·4$$

The equivalent weight of copper $= 31 \cdot 785$

$\dfrac{68 \cdot 4}{31 \cdot 785} = 2 \cdot 15$, to which the nearest whole number is 2

Hence the atomic weight of copper is $31 \cdot 785 \times 2 = 63 \cdot 57$.

(ii) In 1819 Mitscherlich, a pupil of Berzelius, discovered that compounds of similar chemical composition formed similarly shaped crystals. For example, he found that potassium hydrogen phosphate and potassium hydrogen arsenate have the same crystalline form and are composed of the same number of atoms, only differing in that 75 parts of arsenic in one take the place of 31 parts of phosphorus in the other. The shape of the crystal reflects the arrangement of the atoms in the unit cells of which it is composed, and in these 'isomorphous' substances, the atoms of one element replace those of another, one for one, without altering the shape of the crystal (see Chap. 4).

Mitscherlich and Berzelius used the Law of Isomorphism to obtain the atomic weights of a number of elements. Thus, since 31 is the atomic weight of phosphorus, the law indicates that 75 is the atomic weight of arsenic. This confirms the value obtained in other ways. Arsenic, antimony and bismuth form some isomorphous compounds, whence the atomic weights of antimony and bismuth can be obtained. In 1828 Mitscherlich showed that potassium selenate was isomorphous with potassium sulphate. The formula of potassium sulphate can be found from a knowledge of its percentage composition and the atomic weights of potassium, sulphur and oxygen. The law of isomorphism indicates that potassium selenate will have an analogous formula, viz. K_2SeO_4. Hence, from the composition of this compound, the atomic weight of selenium can be calculated.

Mitscherlich found (1832) that potassium permanganate and perchlorate were isomorphous, and that potassium manganate and chromate were isomorphous with the sulphate, whence the atomic weights of manganese and chromium could be obtained. He had also demonstrated earlier the isomorphism of the simple and double sulphates of calcium, magnesium, manganese(II), iron(II), chromium, zinc, cobalt, and nickel, and of the crystalline minerals in which aluminium, chromium and iron occur. In 1828 Berzelius published a new table of atomic weights, in which many of the previously accepted values for the metals were revised as a result of Mitscherlich's observations, and which led to a change in the accepted formulae for their compounds.

(iii) This century has seen the development of the mass spectrograph, an instrument capable of comparing the masses of atoms to an accuracy of at least 0·01%. It was first made by F. W. Aston about 1919, and was based on the discovery by J. J. Thomson in 1912 that the positively charged particles produced in a discharge tube containing a gas at a suitable low pressure could be separated according to their masses. This was done by deflecting them by a magnetic field and focusing them by an electric field onto a photographic film. From the relative positions of the spots caused by different atoms, their masses could be compared.

It was found that the masses of all the atoms of certain elements were not all the same. Atoms of a given element having differing masses are called 'isotopes'. For example oxygen exists as three isotopes having masses of 16, 17, and 18 in the proportions 99·76%, 0·04% and 0·20% respectively. In taking $O = 16$ as the basis for a scale of atomic weights, values are obtained which therefore differ slightly from those calculated from measurements relative to the mass of the isotope ^{16}O. There is now international agreement to refer atomic weights to a unit equal to one twelfth of the mass of the carbon isotope ^{12}C.

A modern mass spectrometer is not only capable of measuring the masses of atoms with great accuracy, but can also be used as an analytical instrument for detecting the nature both of atoms and groups of atoms in a minute quantity of an unknown substance.

Formulae

The concept of a molecule arose from a study of the behaviour of gases. It is clear from the foregoing that the molecular formula of a gas or vapour can be determined from a knowledge of its composition, the atomic weight of its constituents and its relative density. For example, a certain liquid is shown to consist of 36 parts by weight of carbon and 7 parts of hydrogen. Its empirical formula is therefore C_3H_7. The relative density of the vapour is 43, hence the molecular weight is 86, and the formula of the molecule is thus C_6H_{14}.

The molecular weight and the molecular formula apply to the substance when in the gaseous state. When a gas condenses to a liquid or solid, the concept of a molecule becomes less clear.

In a crystalline solid, the atoms are arranged in a pattern which extends continuously throughout the crystal grain. X-ray studies show that the repeating unit that makes up the crystal of a substance such as dinitrobenzene consists of the group of atoms $C_6H_4(NO_2)_2$, and hence the entity of the molecule persists in the solid state. But in the case of an ionic compound such as sodium chloride, where the sodium and chlorine ions occupy alternate positions in a cubic lattice, a particular sodium ion is associated no more closely with one neighbouring chlorine ion than with another and the term 'molecule' is inappropriate. The terms 'macromolecule' or 'giant structure' are used.

A liquid may be regarded either as a condensed gas or a molten solid, the latter being the more fruitful point of view. On the former view, the liquid may consist of a close mixture of molecules as they existed in the gas phase or of groups formed by some kind of bonding between two or more molecules. When a solid melts, the orderly arrangement present in the crystal may entirely give way to disorder and form a random mixture of particles, or the liquid may retain, to a greater or less degree, some of the 'order' present in the crystal. This view is particularly successful in explaining the physical properties of water (see p. 92). In Chapter 6 we shall see that the terms 'molecule' and 'molecular weight' can be assigned a clearer meaning when applied to substances in solution.

Many formulae in common use should therefore be regarded, when applied to solids or liquids, as 'empirical' formulae rather than molecular formulae.

The mole

The word 'gram-molecule' denotes a certain quantity of the compound, i.e. the molecular weight expressed in grams. Similarly, the words 'gram-atom' and 'gram-ion' are used and have corresponding meanings. The number of molecules in a gram-molecule is called the Avogadro Number, and has a value of about 6×10^{23}. It is also the number of atoms in a gram-atom and of ions in a gram-ion. A useful word, the 'mole', is now used to mean an Avogadro Number of molecules, atoms or ions, i.e. one speaks of a mole of chlorine, or of copper(II) ions or of copper sulphate crystals, meaning 71 grams, 63·57 grams or 250 grams respectively. It should be noted that although the word 'mole' is derived from 'molecule' it is applied nowadays to a 'formula weight', whether it be of an atom, ion, radical, molecule or empirical formula such as NaCl.

(c) Chemical bonding

The nature of the forces between the atoms constituting a chemical compound has long been a matter for speculation. Until something was known about the nature of the atoms themselves, no serious theory of bonding was possible. However, before atomic structures were elucidated, it was realized by chemists that different types of chemical bonding exist. For example, the forces between the atoms in the radicals —SO_4 and —NO_3 are evidently stronger than those between these radicals and the metals with which they form salts. These latter bonds, in turn, are of a different kind from those in complex salts, for the 'combining capacities' of potassium and of the cyanide radical, and of iron(III) and the cyanide radical, appear to be mutually satisfied in the compounds potassium cyanide and iron(III) cyanide respectively, and yet there must be some 'residual bonding force' that holds these two compounds together in the complex salt potassium ferricyanide.

The discovery of the existence of particles lighter than the lightest atom opened up the possibility that atoms could have a structure. This is not the place to trace the evolution of our ideas on atomic structure; reference should be made to other textbooks (see Bibliography). It was clear from the Rutherford-Bohr theory that chemical interaction between atoms involves the outer shells of electrons. The most stable and unreactive elements, the inert gases, have an atomic structure in which the outer shell consists of eight electrons (two in the case of helium). On the assumption that, in chemical interaction, atoms would tend to acquire an inert-gas structure, Kossel and G. N. Lewis independently put forward the view in 1916 that one form of chemical linking consists in the electrostatic attraction between oppositely charged atoms, formed as a result of the transfer of one or more electrons from one atom to the other, resulting in the production of inert-gas structures. Thus, the bonds in sodium chloride, for example, and in most salts and strong acids and bases, are of this kind and are called electrovalent bonds. The formulae of such compounds may be written:

$$\mathrm{Na^+\left[:\overset{..}{\underset{..}{Cl}}:\right]^-}, \qquad \mathrm{Ca^{++}\left[:\overset{..}{\underset{..}{Cl}}:\right]_2^-}, \qquad \mathrm{Mg^{++}\left[:\overset{..}{\underset{..}{O}}:\right]^=}.$$

Sodium chloride   Calcium chloride   Magnesium oxide

As we shall see in later chapters, such substances exist in the solid state as an aggregate of charged particles or 'ions', and not of neutral atoms, and are referred to as 'ionic' substances.

The electrovalent bond is not adequate to explain all types of chemical bond, for it is clear that the forces holding together the atoms in, say, ammonia or methane, are not of this kind, for these compounds do not ionize into hydrogen and nitrogen or hydrogen and carbon ions. In 1916 G. N. Lewis also introduced the conception of the 'covalent bond', formed as the result of the sharing of two electrons between two atoms. One pair of shared electrons constitutes a single covalent bond. The bonds in non-electrolytes, including, for example, nearly all organic compounds, are of this kind. A covalent bond is represented by the ordinary valency bond symbol —, and, unlike the electrovalent bond, has a definite direction in space. The following formulae of covalent compounds may be quoted as examples:

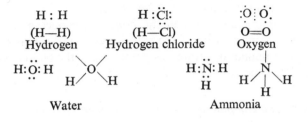

Both electrons in the shared pair may be contributed by the same atom: in this case, the bond is known as a 'co-ordinate covalency'. For example, the unshared pair of electrons in the ammonia molecule may be donated to an atom that is two electrons short of its inert-gas structure, with the formation of a co-ordination compound. Aluminium chloride forms such a co-ordination compound with ammonia and its formula is

$$
\begin{array}{cc}
\text{Cl} & \text{H} \\
\text{Cl:}\overset{\cdot\cdot}{\text{A}} \leftarrow & \text{:}\overset{\cdot\cdot}{\text{N}}\text{:H} \\
\text{Cl} & \overset{\cdot\cdot}{\text{H}}
\end{array}
$$

(see Expt. 1c-1).

The electrovalent bond and the covalent bond may be regarded as two extreme forms of linkage, and transition from one to the other is sometimes possible. For example, the hydrogen in the molecule of an acid may be held by a covalent bond which becomes electrovalent when the acid ionizes. Expt. 1c-2 below shows that hydrogen chloride exists in the un-ionized form H—Cl when dissolved in toluene, but ionizes when dissolved in water. It is known that hydrogen ions do not exist as such in water, but are co-ordinated to water molecules forming H_3O^+ ions. Thus, when hydrogen chloride gas dissolves in water, the reaction that occurs may be represented by the equation

$$HCl + H_2O \rightarrow H_3O^+ + Cl^-.$$

The electron distribution between the two atoms joined by a covalent bond may not be uniform, and so a range of intermediate conditions may exist between that in which the valency electron is shared equally by the two atoms and that in which it is transferred completely to one atom, forming an ion. The non-uniform distribution of the valency electrons leads to the formation of a more or less 'polar' molecule possessing a 'dipole moment' (see p. 92). In the extreme case, the 'dipole' becomes a pair of oppositely charged ions, and the 'polar' substance is termed 'ionic' (see Expt. 1c-3).

Even in compounds in which the bonds were once assumed to be wholly ionic, some sharing of the transferred electron may occur. The following compounds illustrate this point: in LiF the bond is 84% ionic, in LiH, 77% ionic, and in HF, 39% ionic.

Expt. 1c-1   The compound of ammonia with aluminium chloride

Prepare about 5 g of aluminium chloride by passing dry chlorine over heated aluminium turnings and collecting the product in a wide-mouthed bottle in the usual manner. Pass a stream of ammonia, dried by passage through a tower of quicklime, into the bottle containing the aluminium chloride. Much heat is generated and a liquid forms which, on cooling, solidifies to a waxy solid. Transfer this material to a hard-glass test-tube fitted with an air condenser about 1 m long. When the tube is heated strongly, the co-ordination compound sublimes and collects in the condenser. It may be removed when cold and will be found to consist of a stable, white solid that does not fume in damp air, as aluminium chloride itself does. The composition of the compound may be verified by determining the nitrogen content by Kjeldahl's method. (See, for example, Mann and Saunders, *Practical Organic Chemistry*.)

Expt. 1c-2   The reaction between hydrogen chloride and water

Dry about 30 cm³ of toluene by allowing it to stand in contact with anhydrous calcium chloride in a closed flask for a few minutes. Pass a stream of hydrogen chloride gas, dried by concentrated sulphuric acid, into the dry toluene. Use this solution for the following tests:

(i) Test the solution with litmus paper.

(ii) To a small quantity add a few marble chips.

(iii) Immerse two copper wires in the solution and connect them through an ammeter or a torch bulb to a 4 V battery. Note that the solution does not conduct electricity.

Now shake the remains of the solution with an equal volume of distilled water in a separating funnel. Run off the aqueous layer. Test small portions (i) with litmus paper, (ii) with marble chips, (iii) for electrical conductance, (iv) with silver nitrate solution. It is clear that some, at least, of the hydrogen chloride has passed into the aqueous layer, and that, in so doing, it has developed different properties.

Finally, shake the remainder of the aqueous layer with an equal volume of toluene. Separate the layers and shake the toluene with a further quantity of distilled water. Test this latter with silver nitrate solution. No precipitate forms, showing that hydrogen chloride cannot pass from an aqueous solution into toluene, the reason being that, once the hydrogen chloride enters the water, it no longer exists as H—Cl, but as H_3O^+ and Cl^- ions, which do not dissolve in toluene.

Expt. 1c-3 **Polar molecules**

When there is unequal sharing of electrons between atoms in individual molecules, parts of the molecule will be slightly positive and parts will be slightly negative with respect to the rest. If a gas consisting of such molecules is placed in the electric field between the charged plates of a condenser, the polarized molecules will orientate themselves with the result that the potential gradient between the plates will be reduced. The capacity of the condenser formed by the plates is thus changed.

In this experiment, a cylindrical condenser, through which a gas can be caused to flow, forms part of an oscillating circuit working at a low radio-frequency and any change in the capacity of the condenser manifests itself as a change of note.

A simple cylindrical condenser can be made from two brass tubes, one about 1 cm in diameter and the other big enough to slide outside it leaving a gap of 1–2 mm. The inner tube is fitted with rubber rings at each end and these fit into the outer tube. Two side arms are brazed on to the outer tube to enable gas to be passed through the annular space between the two tubes.

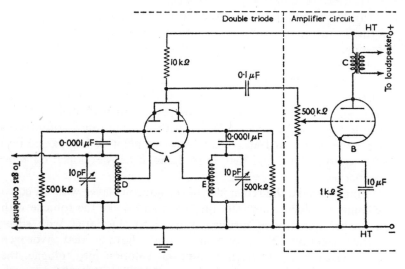

A   Double triode, e.g. ECC81
B   Triode, e.g. 6JSG
C   Output transformer, e.g. 30 : 1
D,E Centre-tapped coils of 500–1000 turns of cotton-covered wire about 40 S.W.G.
(The magnitudes are unimportant: those quoted above give a satisfactory circuit)

Fig. 3

In practice it is best to construct two r.f. circuits beating with an audible frequency and to modify the latter by means of the gas condenser. To do this construct the circuit shown in the diagram. There are two oscillators working in a double triode for convenience. The output from this is fed into an amplifier triode through a volume control and the final output goes to a loudspeaker. It will be found easy to get an audible note by tuning with the variable condensers. To avoid effects from external charges, the whole apparatus should be surrounded by a metal screening.

Connect the gas condenser with the circuit and blow dry air or dry hydrogen through it. No change of note should be heard. Now blow dry hydrogen chloride through the condenser. A very marked change, which may take the frequency outside the audible range, will occur. Blow dry air again: and restore the note.

Try the effect of blowing the following gases through the condenser:

(1) Damp air
(2) Dry hydrogen bromide
(3) Dry hydrogen iodide
(4) Dry sulphur dioxide
(5) Dry ammonia

What indications are there about the structure of their molecules?

THE STRUCTURE OF MOLECULES

It is now possible to get a good deal of information about the structure of molecules from a study of spectra. Radiation of wavelengths from 1 to 10^{-7} m is absorbed by gases and liquids and energizes the molecules in different ways. For example, microwave radiation (wavelengths from 1 to 10^{-3} m) affects the rotation of molecules, i.e. the frequency in this waveband corresponds to energy quanta of a magnitude associated with the rotational energy of simple molecules. By measuring the wavelengths absorbed it is possible to deduce the moments of inertia of simple molecules, and sometimes the lengths of the bonds between the atoms constituting the molecule and the angle between them.

Again, radiation in the infra-red region (wavelengths from 10^{-4} to 10^{-6} m) affects the vibration of various parts of the molecules, i.e. can energize the bonds between the atoms in the molecules. The study of the infra-red absorption spectrum of a liquid can thus provide information about the flexibility of the bonds between the atoms in the molecules. By measuring the wavelength of the absorption bands in the spectrum, various bond strengths can be compared. But perhaps the most important use of infra-red spectra is in analysis: by discovering what functional groups are present in the molecule, the compound can be identified. Infra-red absorption spectra are characteristic of the molecule and infra-red spectroscopy is an indispensable aid to analytical and synthetic chemistry. Small infra-red spectroscopes are now available at low cost and the infra-red spectra of liquids and solutes can be easily plotted in the school or college laboratory. By choosing a range of substances with known molecular structures, such as a series of alcohols, the student will be able to pick out the infra-red absorption bands corresponding to certain groups, in the above example, the hydroxyl group. Reference to more specialized books will enable the infra-red spectra of more complex substances to be identified.

(d)　Radiochemistry

LABORATORY RULES WHEN USING RADIOACTIVE SUBSTANCES

The health hazard from radioactivity when doing these experiments is negligible. Nevertheless, it is wise to take precautions in order to get accustomed to the procedures necessary when using radioactive materials in general.

(1) For substances of such low activity as are used in the experiments described here, a special laboratory is not required, but a special section of the bench should be set aside for them. The work should be carried out on a tray lined with polythene sheet or absorbent paper.
(2) Hands should be well washed after an experiment.
(3) Solutions should be added by means of teat pipettes—never put a pipette, or anything else, to the mouth.
(4) Waste should go into marked receptacles.
(5) All apparatus should be marked and kept for use with radioactive substances.
(6) Radioactive substances should never be touched by hand and should not be used if there is a cut or other wound on the hands.
(7) If a spill occurs, mop up with absorbent tissue, wash liberally and monitor when dry.

Expt. 1d-1

Place a key or other metal object on a photographic plate or piece of film (previously wrapped in black paper in the darkroom) and place it in a shallow dish. Cover the key and wrapped film with powdered uranium or thorium salt and leave it in a cupboard for several days. Remove the plate or film in the darkroom and develop it.

In 1896 Becquerel discovered that certain uranium minerals emitted rays which could pass through black paper, affect a photographic plate, and ionize the air. Madame Curie showed that thorium compounds also emitted these rays. This phenomenon became known as 'radio-activity'. Two years later, Madame Curie isolated the salt of a very much more radioactive element, subsequently called radium. The radiations were shown to be of three kinds: α-rays, which consist of positively charged helium atoms; β-rays, which are fast-moving electrons; and γ-rays, which are very penetrating X-rays, i.e. a form of electromagnetic radiation. It was clear that the atoms of the elements in these minerals, i.e. those of radium, uranium, thorium, are losing particles that have both charge and mass, and so the atoms of these radioactive elements must be decreasing in mass, changing their charge and forming atoms of other elements. This theory of radioactive disintegration was first put forward by Rutherford and Soddy in 1903. These processes of radioactive transformation have been widely studied and the following are examples of radioactive decay series.

1.

Atomic number	90	88	90	88	86	84	82	83	82
Element	Th	Ra	Th	Ra	Rn	Po	Pb	Bi	Pb
Old name	Th	MsTh	RaTh	ThX	Tn	ThA	ThB	ThC	ThD
Atomic mass	232	228	228	224	220	216	212	212	208
Rays emitted	α	2β	α	α	α	α	βγ	αβ	—
Half-life	1.4×10^{10} years	6·7 years	1·9 years	3·6 days	52 secs	0·16 secs	10·6 hrs	60·5 mins	∞

2.

Atomic number	92	90	91	92	88	...	82
Element	U	Th	Pa	U	Ra	...	Pb
Atomic mass	238	234	234	234	226	...	206
Rays emitted	α	βγ	δγ	2α	5α,4βδ		—
Half-life	4.5×10^9 years	24·1 days	1·17 mins	2.5×10^5 years	1500 years	...	∞

It is interesting to measure the rate of decay of a suitable species, and such an experiment is described below. As will be seen by reference to the decay curve formed, it follows a 'first order reaction', i.e. the time needed for half the atoms to decay is the same no matter how many atoms there were at the start. This time is known as the half-life, and the half-lives of some radioactive elements are shown in the decay series above.

The following experiments illustrate how the product of decay of a radioactive element may be separated chemically. This product is often radioactive itself and is decaying to other radioactive elements, and so on until the end of the series is reached. In a sample of a radioactive salt—for example, thorium nitrate—there is eventually an equilibrium between the various decay products, although this may take many years to establish owing to the long half-life and slow rate of growth of some of the radioactive elements. If one interferes with this by separating some of the constituents chemically, or by ion exchange or solvent extraction, the elements then decay or grow in such a way as to tend to restore the radiochemical equilibrium. Examples of the study of such decay and growth curves are given in the experiments. For the first four experiments, a Geiger-Müller (G.M.) counter for use with liquids is needed. It excludes α-rays but will count β-rays.

In part (1) of Expt. 1d-2, a decay product of thorium, i.e. the radio-active lead-212, is separated from thorium by precipitation as hydroxide. Because only very small amounts are present, some lead nitrate solution is added as a 'carrier' and this, when precipitated as hydroxide, brings down the lead-212 with it. In part (2) of this experiment, another decay product of the thorium, i.e. radium-224, is separated by precipitation as sulphate, barium nitrate having been added as a 'carrier'. This then decays into lead-212, whose growth curve can therefore be obtained. For this it is necessary to use a counter which excludes the α-rays from the radium and detects the βγ-rays from the lead-212.

In Expt. 1d-3, the decay curve of another product from thorium, i.e. bismuth-212, is plotted. This is separated from the lead-212 by precipitating the latter as sulphate. The bismuth-212 is then precipitated as sulphide, some ordinary bismuth salt having been added to act as a carrier. The bismuth-212 has a shorter half-life (about 1 hour) than the lead-212 (about 10 hours), so the decay curve can be determined more conveniently for bismuth-212 than for lead-212.

Expt. 1d-4 uses a different means of separating the radioactive products, i.e. solvent extraction. The example is taken from the uranium decay series (see p. 25). Uranium-238 decays via thorium-234 and protoactinium-234 to uranium-234. The protoactinium has a half-life of only just over 1 minute, so its decay curve is quickly plotted. It is very easily separated from the uranium and thorium because its salts are soluble in organic solvents.

The use of a 'radioactive tracer' is illustrated by Expt. 1d-5. The method is used to show that when a solute is in equilibrium with its saturated solution, the equilibrium is 'dynamic', i.e. solid particles are passing into solution and solute particles are going back into the solid state. Lead chloride is chosen as the solute and is 'labelled' by the addition of a thorium salt which contains the radioactive lead-212. When this is shaken with a solution that is already saturated with lead chloride, the solution becomes radioactive, showing that lead-212 has passed into the solution. This can only have been by exchange with lead ions already in the solution.

Lastly, the photographic emulsion method of studying radioactive decay processes is illustrated in Expt. 1d-6. The α-particles emitted by radioactive substances affect a photographic plate, and this fact has been much used from the earliest days in the study of radioactivity. When the plate is developed, the tracks become visible. The length of the track produced by a particle is a measure of its energy. Once these are known, a comparison of the lengths of tracks enables the element emitting the α-particle to be identified.

Expt. 1d-2   The decay and growth curves of thorium B (lead-212)
(The two parts of this experiment can be run concurrently.)

Prepare a solution of thorium nitrate by dissolving about 1 g in 20 to 25 cm³ of distilled water. To about 10 cm³ of this solution add three drops each of approximately 5% solutions of lead nitrate and barium nitrate.

(1) Add 0·880 ammonia dropwise until precipitation of hydroxides is complete. This precipitate consists of thorium and lead hydroxides, the latter carrying the radioactive isotope lead-212, which was contained in the original thorium nitrate. The radium isotope, radium-224, which is also a product of decay of the thorium (see p. 25), remains in solution in the filtrate, the barium ion acting as a 'hold back' carrier for it. The precipitate of thorium and lead hydroxides should be warmed to coagulate it and centrifuged. Decant and preserve the liquid for part (2). Dissolve the precipitate in the least possible quantity of nitric acid and re-precipitate with 0·880 ammonia. Centrifuge this, decant the clear liquid and again dissolve the precipitate in nitric acid and re-precipitate with 0·880 ammonia. Reject the washings in the latter two cases. The object of this is to free the precipitate from all adsorbed ions which may have been carried down when it was first precipitated. Now wash the precipitate with hot water, centrifuge, and dissolve it in about 5 cm³ of dilute nitric acid. Place this in the liquid counter, and take a count over a period of, say, two minutes repeating this every hour or at convenient intervals during the next twenty-four hours. The count is due to β-particles emitted by the lead-212, the β-particles from the radium-228 being of too low an energy to affect the G.M. tube. Plot the decay curve for the lead-212. The half-life is about 10·6 hours.

(2) This part consists in plotting the growth of the same isotope, namely lead-212, from radium-224, itself a decay product of thorium. The radium, however, was not precipitated by the ammonia, but remained in the filtrate which was preserved. Treat this filtrate (which contains the barium added to the original solution), with dilute sulphuric acid until no further precipitation occurs. This precipitate of barium sulphate will carry down with it the radium sulphate. (It will also bring down radium-228, if present, and thus the

growth curve may include actinium-228 produced from the radium-228. Whether or not this happens to any appreciable extent depends upon the age of the thorium salt.) Centrifuge the precipitate, decant the liquid, transfer the precipitate to a watch glass and dry it over a beaker of boiling water. Stick a piece of sellotape over the precipitate. This not only retains it mechanically on the watch glass but retains the radon which is a decay product of the radium and in turn decays to lead-212, whose growth curve is being plotted in this part of the experiment.

Invert the watch glass over the G.M. tube and count at convenient intervals over a period of twenty-four hours.

It should be recalled that the G.M. counter for use with liquids has a glass shield which cuts out α-particles but is transparent to β-particles, so that in this experiment it is possible to measure the growth of the lead-212 which is a β-emitter in spite of the presence of radium-224 and radon-220 which are α-emitters.

Expt. 1d-3   The decay of bismuth-212

In this experiment we shall measure the decay curve of the next isotope in the series, namely bismuth-212, which is formed from the lead-212 studied in the previous experiment as a result of β-emission. This is a slightly easier decay curve to plot because the half-life of the bismuth-212 is about one hour rather than ten hours. Dissolve about 1g of thorium nitrate in 20 to 25 cm^3 of distilled water. To this solution add a few drops of lead nitrate solution (approx. 5%) and a few drops of bismuth nitrate solution. The latter acts as a 'hold back' carrier for the bismuth-212 at this stage. Add dilute sulphuric acid dropwise to precipitate the lead, centrifuge, and to ensure that all the lead is removed from the solution, repeat the precipitation by adding a few more drops of lead nitrate to the filtrate and adding dilute sulphuric acid again. Centrifuge off the precipitate. Pass hydrogen sulphide into the clear solution to precipitate the bismuth as sulphide, centrifuge, reject the clear liquid. Dissolve the bismuth sulphide, which includes the bismuth-212, in a few drops of concentrated nitric acid, dilute with 2 or 3 cm^3 of distilled water and pour into the G.M. tube. Count at once for a minute and repeat the count every five minutes for about half an hour and then every ten minutes for a further half hour. Continue to count at convenient intervals for another three hours or so. Plot the decay curve of the bismuth-212. The half-life should be 60 minutes.

Specimen result

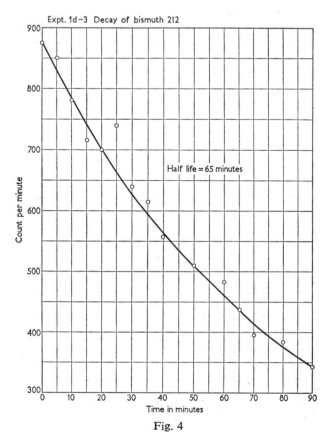

Fig. 4

Expt. 1d-4   The half-life of protoactinium

Dissolve about 1 g of uranyl nitrate in about 3 cm³ of distilled water. Add 7 cm³ of concentrated hydrochloric acid, transfer to a separating funnel and shake with about 10 cm³ of amyl acetate for five minutes. (Several other organic solvents are equally good.) Run off the aqueous layer and transfer 5 cm³ of the amyl acetate layer to the liquid counter. At once count for 15-sec periods at 5-sec intervals and plot a graph. The half-life should be about 1·2 minutes.

A growth curve for protoactinium can be obtained by counting the aqueous layer. The layers may then be shaken together again and the experiment repeated.

Specimen result

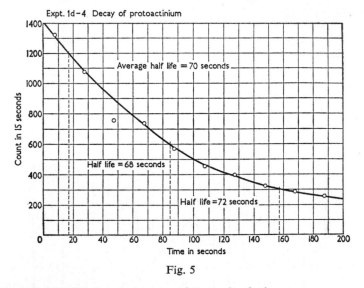

Fig. 5

Expt. 1d-5 Equilibrium between solute and solution

The equilibrium between solid lead chloride and its solution is followed by a radioactive tracer.

Make a saturated solution of lead chloride by adding dilute hydrochloric acid to about 15 cm³ of a solution of lead nitrate in a test-tube. Shake well, centrifuge off the solid, transfer 5 cm³ of the solution to the G.M. tube and count it over a period of one minute. It should give no more than a background count. To about 10 cm³ of the stock solution of thorium nitrate add a few drops of lead nitrate and precipitate lead chloride by the addition of hydrochloric acid. This precipitate will contain the radioactive isotope lead-212. Centrifuge the precipitate and wash it thoroughly three times with the distilled water. Add it to the saturated lead chloride solution prepared above and shake or stir mechanically for about twenty minutes. Centrifuge off the precipitate and count the clear liquid over a period of one minute. The liquid will now be active, due, not to the solution of more lead chloride (for it was already saturated), but to the interchange of lead ions between the solution and the solid as indicated by the radioactive tracer. The count rate will be found to be at least four or five times greater than that obtained with the original saturated lead chloride solution.

Expt. 1d-6   Photographic recording of α-particle tracks

Some 75 mm × 25 mm nuclear research photographic plates (C_2 or G_5) are required. In a darkroom, soak one of the plates for 10 min in a 1% solution of a thorium salt, wash with distilled water and leave to dry in the dark for a few days. Develop it according to the instructions supplied with the plates, fix thoroughly, wash and dry it. Examine the plate under a microscope, using a 5 mm and then a 4 mm objective. It will be found to be covered with star-like images, each one produced by the tracks of α-particles from radiothorium and its decay products (see illustration). The stars are three-dimensional, but occasionally one may be found with its tracks more or less in the plane of the emulsion. The lengths of the tracks can be compared by the use of an eye-piece with a scale, or by means of a mechanical stage on the microscope.

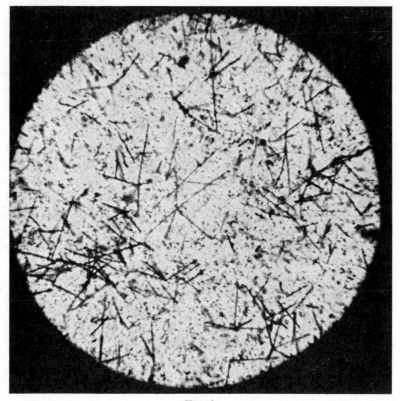

Fig. 6

ENERGIES AND RANGES OF α-PARTICLES FROM
VARIOUS SOURCES

Element	Energy in MeV	Range in Microns
Thorium	5·42	22
Radium-224 (ThX)	5·68	24
Radon-220 (Thoron)	6·28	28
Polonium-216 (Th A)	6·78	32
Bismuth-212 (Th C)	6·04	26·5
Polonium-212 (Th C')	8·78	47
Polonium-212 (ThC')	9·45	53
Polonium-212 (ThC')	10·62	63·5

TWENTIETH CENTURY ALCHEMY

We have seen that a radioactive element, while emitting α-, β-, and
γ-rays, is changing into other elements, i.e. it is undergoing naturally
the sort of process the alchemists were trying to bring about by their
attempts to change lead into gold. It can be said that Rutherford was
the first successful modern alchemist when he showed, in 1919, that
high energy α-particles reacted with nitrogen atoms to form hydrogen
and an isotope of oxygen:

$$^{14}N + {}^4He \rightarrow {}^{17}O + {}^1H$$

Since then, many other such 'alchemical' reactions have been brought
about. In these reactions, one 'reagent' always consisted of missiles
from a naturally radioactive element. The first fully artificial alchemical
change was achieved by Cockcroft and Walton in 1933, when they
speeded up protons in a 1-million volt field and directed them on to a
lithium compound. They were able to show that an occasional lithium
atom was hit, giving rise to two α-particles and a good deal of energy:

$$^7Li + {}^1H \rightarrow {}^4He + {}^4He$$

Among other changes of a similar kind that were discovered soon
after, were:

$$^{14}N + {}^1H \rightarrow {}^{11}C + {}^4He$$

in which an isotope of carbon is produced, and

$$^9Be + {}^4He \rightarrow {}^{12}C + {}^1n$$

The latter equation represents the reaction whereby Chadwick, in
1932, discovered the neutron. Consisting only of a nucleus with no

electrons, a neutron is capable of penetrating other atoms, reacting with the nucleus and bringing about a great many alchemical changes, e.g.

$$^{10}\text{B} + {}^1\text{n} \longrightarrow {}^7\text{Li} + {}^4\text{He}$$

$$^{27}\text{Al} + {}^1\text{n} \longrightarrow {}^{24}\text{Na} + {}^4\text{He}$$

$$^{27}\text{Al} + {}^1\text{n} \longrightarrow {}^{27}\text{Mg} + {}^1\text{H}$$

$$^{27}\text{Al} + {}^1\text{n} \longrightarrow {}^{28}\text{Al} + \gamma$$

When aluminium is bombarded by neutrons, the speed of the neutrons determines which of the three reactions occurs, fast neutrons producing the first, and slow neutrons the last reaction. The latter illustrates a type of reaction in which the neutron is 'captured', no other particle being formed. This capture reaction is important when the heavy elements are concerned. Thus, in an atomic pile, slow neutrons are captured by uranium atoms, ^{235}U, leading to fission of the atoms, in which an average of 2·5 fresh neutrons are produced for every neutron captured. If an average of more than one of the fresh neutrons undergoes capture, a 'chain reaction' ensues. The neutrons can also be captured by the structural materials of the pile, so its geometry can be arranged to prevent the occurrence of a chain reaction. In addition, uranium-238 atoms also capture neutrons leading to the formation of plutonium-239. The latter has a half-life of 2·1 $\times$ 10^4 years. The overall reaction is

$$^{238}\text{U} + {}^1\text{n} \longrightarrow {}^{239}\text{Pu} + 2\beta$$

Artificial radioactivity

Curie-Joliot (1934) discovered that some products of these 'alchemical' reactions were radioactive. This artificially-induced radioactivity has led to the production of such substances as radio-phosphorus and radio-sodium, which have been of great value in several branches of science. Radioactive materials of this sort have a short half-life and are useful, for example, as radioactive 'tracers', in the study of the mechanism of chemical reactions and of biological processes, and in replacing radium in the treatment of cancer.

Examples of reactions in which artificially induced radioactivity is produced, are:

$$^{19}\text{F} + {}^4\text{He} \longrightarrow {}^{22}\text{Na} + {}^1\text{n} \longrightarrow {}^{22}\text{Ne} + e^+$$
$$\text{(half-life 6 months)}$$
$$^{25}\text{Mg} + {}^4\text{He} \longrightarrow {}^{28}\text{Al} + {}^1\text{H} \longrightarrow {}^{28}\text{Si} + e^-$$
$$\text{(half-life 2 minutes)}$$

B C—D

2 The gaseous state

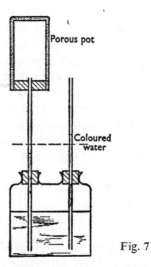

Fig. 7

Set up the apparatus shown in Fig. 7. Displace the air from a large beaker with coal-gas or hydrogen, covering the mouth of the beaker with a card. Invert the beaker, remove the card and quickly place the beaker over the porous pot attached to the double-necked bottle. Note and interpret the behaviour of the coloured liquid in the tubes. Remove the beaker and again note what happens to the liquid. Repeat the experiment, using carbon dioxide in the beaker instead of hydrogen.

Expt. 2-2

Fill a gas jar with nitrogen tetroxide or sulphur dioxide and invert it over a jar of air. After a few seconds, test for the presence of the heavier gas in the lower jar. Repeat the experiment, but place the jar of air above the jar containing the heavy gas. Test for the presence of the latter in the upper jar after (a) a few seconds, (b) a few minutes.

Expt. 2-3

Seal a little bromine in a thin glass bulb. Place it in a glass bottle fitted with a rubber stopper carrying a tap and a vaselined brass rod with which the bulb can be broken. Compare the rates of diffusion of the vapour through the bottle when (a) it is full of air, (b) the air has been pumped out.

Expt. 2-4 Molecular motion in mercury vapour

Prepare a Pyrex glass tube about 20 cm long and 1 cm external diameter, with one end closed and the other drawn to a thick-walled constriction to be sealed off later. Place about 5 cm³ of mercury in the tube and cover this with about ½ cm of crushed glass of about 1 mm particle size (no glass dust should be included). Clamp the tube vertically and evacuate it by means of a good vacuum pump. Warm the tube with a Bunsen burner and boil the mercury *in vacuo* for a few minutes. Then seal off the tube at the constriction. Boil the mercury and note that the glass particles gather in a cloud well above the mercury surface, buoyed up by bombardment by the mercury atoms.

THE GASEOUS STATE AND THE GAS LAWS

The above experiments illustrate one of the most striking characteristics of gases, namely, their power of diffusion. The ability of gases to mix freely and rapidly with each other, and to pass readily through many solid substances, strongly suggests that view of their inner structure which is now called the Kinetic-Molecular Theory. This section reviews the facts upon which the theory is based and recalls the historic experiments and reasoning of Boyle and his successors.

The gaseous state was first formally recognized early in the seventeenth century by van Helmont, who introduced the name 'gas'. The Hon. Robert Boyle made the first quantitative studies of the behaviour of gases (1661). Having observed the distension of a lamb's bladder when placed under the receiver of an air-pump, he measured the changes in volume of a sample of air confined in a closed limb of a U-tube when mercury was poured into the other limb. When the volume of the enclosed air had been reduced to half its value at atmospheric pressure, he 'observed, not without delight and satisfaction', that its pressure was then 2 atmospheres. The results of his measurements are summarized in Boyle's law: 'The volume of a certain mass of gas at a constant temperature is inversely proportional to its pressure.'

The effect of heat on the volume of a given mass of gas was observed by Priestley, Cavendish and Charles, and, about 15 years later (1802), by Dalton and Gay Lussac. Priestley concluded from 'a very coarse experiment' that 'fixed and common air expand alike with the same degree of heat'. Gay Lussac published the relation, known alternatively as 'Charles's law' or 'Gay Lussac's law': 'The volume of a certain mass of gas at constant pressure increases linearly with the temperature, and the increase of a given volume per degree rise in temperature is the same for all gases.' The discovery embodied in the latter part of the statement, i.e. that all gases have the same coefficient of expansion, has been of far-reaching consequence and is the basis of the absolute scale of temperature.

THE KINETIC THEORY OF GASES

Sir Isaac Newton (1680) believed in an atomic theory of matter (see Chap. 1) and held that the atoms in the gaseous state were widely spaced from each other. He attributed the pressure exerted by a gas to the mutual repulsion of its constituent particles. On reducing the space containing the gas, the particles were forced nearer together, the force of repulsion increased and an increase in pressure was thereby produced. The kinetic-molecular theory of gases, which was developed largely from the ideas of Bernoulli (1738), pictures a gas as consisting of widely separated particles of small volume compared with the space occupied by the gas. The particles, to which the term 'molecules' was applied, are in a state of continuous and rapid random motion. The pressure exerted by the gas on the walls of the containing vessel is due to the bombardment of the walls of the vessel by the molecules. Reduction of the volume (at constant temperature) causes an increase in the frequency of bombardment and hence an increase in pressure. If the gas is heated, the heat energy is converted into kinetic energy of the molecules and, if the volume is constant, the pressure consequently rises.

In its simple form, the theory assumes that the molecules are spheres, that they exert no forces on each other, and that their collisions with each other and with the walls of the containing vessel are perfectly elastic. On these assumptions an expression for the pressure exerted by a gas can be derived as follows.

Consider n particles, each of mass m, contained in a cube of side length l. Since the particles are in random motion, there is a normal probability distribution of velocities among them (see p. 40). Consider a particle moving with a velocity c in some arbitrary direction. Its velocity can be resolved into three components u_x, u_y and u_z, normal to the faces of the cube. Then

$$u_x^2 + u_y^2 + u_z^2 = c^2.$$

Each time this particle collides with a wall of the cube normal to the x-axis, its momentum changes from mu_x to $-mu_x$, i.e. there is a change in momentum of $2mu_x$. Before colliding with the same wall again, it must travel a distance of $2l$; therefore it will undergo $u_x/2l$ collisions in unit time. Its rate of change of momentum at the wall is thus $\dfrac{2mu_xu_x}{2l}$.

If c^2 represents a mean value for all the particles, the force exerted by n molecules on one wall is $\dfrac{nmu_x^2}{l}$. The pressure exerted on the wall of the cube is therefore $\dfrac{nmu_x^2}{l^3}$. Since there is no drift of the gas in any particular direction, $u_x = u_y = u_z$, and therefore $u_x^2 = \frac{1}{3}c^2$. If v is the volume of the cube (l^3), the pressure p is given by

$$p = \frac{nm}{v}\tfrac{1}{3}c^2.$$

The total mass of gas, M, is mn, hence

$$p = \frac{1}{3}\frac{M}{v}c^2.$$

The density of the gas, ρ, is equal to M/v, so an alternative expression for the pressure is

$$p = \tfrac{1}{3}\rho c^2.$$

It should be noted that this expression results from the application of the principles of mechanics to the hypotheses of the kinetic-molecular theory. It is therefore very interesting to observe how the relationships between the volume, pressure, and temperature of a given mass of gas observed by Boyle and Charles are consistent with the above equation:

Boyle's law. At constant temperature, the mean kinetic energy of the molecules is constant, i.e. for a given mass, M, of gas, $\frac{1}{2}Mc^2$ is constant. Therefore pv is constant.

Charles's law. At constant pressure, v is proportional to $\frac{1}{2}Mc^2$, so the latter is directly proportional to the absolute temperature.

Graham's Law of Diffusion also follows, for if two gases are at the same temperature and pressure, the rates of free diffusion, assumed to be proportional to the mean velocity of the molecules, will be inversely proportional to the square roots of the densities of the gases.

The kinetic-molecular theory also embraces the important hypothesis of Avogadro, namely, that equal volumes of all gases at the same temperature and pressure contain the same number of molecules. (The truth of Avogadro's hypothesis had been so well established by chemists that Clausius, in 1857, used it as the basis of his proof that temperature is measured by kinetic energy.) For equal volumes of two gases at the same pressure

$$n_1 m_1 c_1^2 = n_2 m_2 c_2^2.$$

It was shown by Clerk Maxwell that, at a given temperature, the mean kinetic energy per molecule in a gas is constant, even if the masses of the molecules differ.

Hence, $\frac{1}{2}m_1 c_1^2 = \frac{1}{2}m_2 c_2^2,$

and therefore $n_1 = n_2.$

Comparison of the densities of gases is thus a comparison of the weights of their molecules.

It is worth emphasizing that the pressure of a certain volume of gas at constant temperature is directly proportional to the number of molecules and is independent of their nature. For a quantity of gas containing n molecules, the gas laws may be written

$$p = \frac{nkT}{v},$$

where k is a constant. For that special quantity of gas called a 'gram-molecule' or 'mole' (i.e. the molecular weight expressed in grams) the constant R is used, and the equation becomes $p = \dfrac{RT}{v}$. For any other mass of gas, say x moles, the pressure will be given by $p = x\dfrac{RT}{v}$. For w grams of gas of molecular weight M, $p = \dfrac{w}{M}\dfrac{RT}{v}$.

An approximate value of R can be obtained from the measurement of the density of a permanent gas and a knowledge of its molecular weight. For example, the density of hydrogen at s.t.p. (273K and 1·013 bar (76 cm of mercury)) is 0·090 grams per litre. Hence 1 mole (2 g) will occupy 2/0·090 litres, and therefore $R = \dfrac{76 \times 13\cdot6 \times 981 \times 200}{273 \times 0\cdot09 \times 4\cdot2 \times 10^7}$, which is approximately 2 calories per mole per deg C.

REAL GASES

The gas laws will only be expected to hold for a gas for which the assumptions of the kinetic theory are true. Such a gas is called an 'ideal gas'. The behaviour of real gases deviates from the gas laws, the deviations being due mainly to the fact that real molecules exert attractive forces on each other, and also that the volume of the molecules of a real gas is finite. As would be expected, the deviations become larger at higher pressures and lower temperatures, when the molecules are closer together. The behaviour of real gases has been expressed by various equations of state, of which one of the most useful is that of van der Waals:

$$\left(p + \frac{a}{v^2}\right)(v - b) = RT.$$

The constant b makes allowance for the volume of the molecules. The existence of attractive forces between the molecules renders the pressure exerted by a real gas less than that for an ideal gas under the same conditions, so a term a/v^2 is added to the measured pressure and represents this 'internal pressure'.

DISTRIBUTION OF ENERGIES AMONG THE MOLECULES

The velocity of an individual molecule is constantly changing owing to collisions; it may momentarily be zero, low, medium or high. The distribution of velocities among the molecules was calculated by Clerk Maxwell from probability laws. It is clear that at any instant most molecules, of course, have a velocity not far removed from the most probable velocity, a few are moving very slowly and a few very fast. The kinetic energy distribution follows the velocity distribution, as the particles are all of the same mass, and the distribution of energy among the molecules, the Boltzmann distribution, is shown in Fig. 8 for one temperature. If the temperature is raised, the number of molecules of higher kinetic energy is increased. (The number having the most probable velocity decreases.) The distribution law may be simplified to the approximate form $n/n_0 = e^{-E/RT}$, where n_0 is the total number of molecules, and n is the total number having an energy greater than any particular value E at the temperature T. This will be referred to later when rates of reaction and energies of activation are discussed (see p. 186).

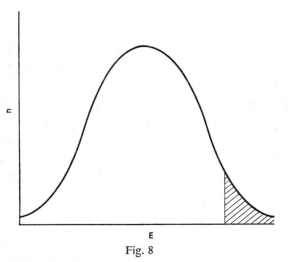

Fig. 8

Expt. 2-5   Gas thermometers

By means of a short length of pressure tubing A attach the bulb, which should have a volume of at least 200 cm³, to a simple manometer made of capillary tubing of diameter about $\frac{1}{2}$ mm and length about 70 cm. Make sure that the bulb and tubing are thoroughly dry. For this purpose, it is advisable to warm the bulb with a Bunsen flame and remove the air with a water-pump at the same time, before joining up at A.

Surround the bulb with a beaker of boiling water until air ceases to bubble out from the manometer tube. Cool the water to about 50 °C, retain it at this temperature for about 5 min and read the height of the manometer. Then cool the bulb in ice and water and, when the pressure has become constant, again read the manometer. Convert the readings to pressures and plot them against the absolute temperatures. The volume of the air in the apparatus is almost

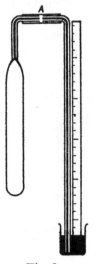

Fig. 9.

constant, as the volume of the manometer is negligible compared with that of the bulb. Find the temperatures of solid carbon dioxide dissolved in industrial spirit or acetone by surrounding the bulb with a Thermos flask containing it, and extrapolating from the graph.

THERMAL DISSOCIATION

It is found that when certain gases (e.g. nitrogen tetroxide) are heated at constant volume the increase in pressure is greater than would be expected if the gas laws were followed. It is clear from the expression for the pressure derived above that one possible explanation is that the number of particles present is increasing. The molecules of nitrogen tetroxide (N_2O_4) tend to dissociate on heating into molecules of nitrogen dioxide (NO_2), so that the hot gas consists of a mixture of double and single molecules in equilibrium, the extent of the dissociation becoming greater as the temperature is raised. The degree of dissociation at any particular temperature may be found as follows.

Suppose that a fraction α of the total number of molecules present, n, has split into two parts. There will then be a total of $n(1 - \alpha) + 2\alpha n = (1 + \alpha)n$ particles. The measured pressure, p_0 will be $(1 + \alpha)$ times as great as the pressure, p_t, to be expected from the gas laws, i.e., $p_0 = (1 + \alpha)p_t$.

It will be noticed that the calculation of p_t requires a knowledge of the molecular weight, M, of the undissociated substance, $p_t = w/M \cdot RT/v$.

In practice, α is usually obtained from the density of the gas, ρ_0, and the density, ρ_t, it would have at the same temperature and pressure if no dissociation had occurred.

$$\rho_t = (1 + \alpha)\rho_0 \quad \text{and thus} \quad \alpha = \frac{\rho_t - \rho_0}{\rho_0}.$$

See Exp. 2–6 below, and p. 257.

Expt. 2-6 Measurement of relative densities of nitrogen tetroxide at various temperatures (Dumas' method)

Prepare six Dumas bulbs by drawing out the end of a 15 cm × 2·5 cm test-tube to a fine neck and bending it over in the flame as shown in the diagram.

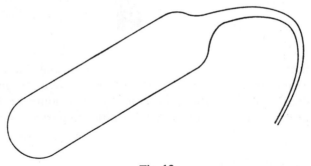

Fig. 10

Dry a test-tubeful of lead nitrate by heating it in a porcelain dish. Transfer it to a hard glass test-tube connected through a U-tube containing dry silica gel to another U-tube immersed in a freezing mixture of ice and salt. Heat the lead nitrate strongly and collect a few cm³ of liquid nitrogen tetroxide.

Weigh the Dumas bulb, warm it in the hand and dip the open end under some liquid nitrogen tetroxide in a crucible. Cool the bulb by wiping it with a cloth wetted with ether. Liquid nitrogen tetroxide will enter the bulb. After about 0·5 cm³ has been drawn in, transfer the bulb to a beaker of warm water, so that the whole bulb is immersed. (It can be weighted down with lead piping or sheet if desired.) Maintain the water at a measured temperature, (say 50 °C) until no liquid nitrogen tetroxide can be seen in the bulb. Seal off the end of the bulb with a mouth blow-pipe. Dry the bulb and weigh it. Then break off the neck under water and measure the volume of water that rushes in. Calculate (i) the weight of this volume of air at the temperature of the experiment; (ii) the weight of the nitrogen tetroxide that had been in the bulb; (iii) the weight of an equal volume of hydrogen at the same temperature. The relative density and apparent molecular weight of the nitrogen tetroxide at this temperature can then be deduced. The degree of dissociation of the N_2O_4 can then be calculated (see p. 42). Repeat the experiment at other temperatures, e.g. 40 °C, 70 °C, 90 °C.

Knowing α, the degree of dissociation, it is possible to calculate K, the equilibrium constant for the dissociation (see Chap. 12). From values of K at several temperatures, the heat change for the reaction can be obtained (see p. 257).

Specimen result

Temperature	49 °C
Weight of bulb empty	12·546 g
„ „ „ with gas	12·613 g
Volume of water entering	41·0 cm³
Weight of this volume of air	0·049 g
Weight of nitrogen peroxide	0·117 g
Weight of an equal volume of hydrogen at 49 °C	0·0031 g
Relative density	37·7
Apparent molecular weight	75·4
Degree of dissociation of N_2O_4 at 49 °C	22%

At 89 °C, the degree of dissociation was found to be 58%.

3 The liquid state

(a) Condensation of gases

Expt. 3a-1

Saturate the air in a large round-bottomed flask with water vapour by swilling it round with water and emptying out the excess. Ignite a small scroll of brown paper, and allow a little smoke to enter the flask. Compress the air by applying the lips to the mouth of the flask and blowing. On suddenly releasing the pressure, cooling occurs and a mist is formed inside the flask. This disappears when the pressure is increased again. If the cycle is repeated several times, improved results will be obtained. The formation of the mist is due to condensation of the water vapour in the flask; the water condenses on the smoke nuclei when the temperature falls owing to the sudden expansion of the air.

Expt. 3a-2

Slowly pass sulphur dioxide from a siphon through a U-tube cooled below $-10°C$ in a freezing-mixture (solid carbon dioxide in industrial spirit is very convenient). Remove the beaker, and note the formation of liquid sulphur dioxide.

The fact that gases cool when allowed to expand from a high pressure to a low, was first noticed by W. Cullen in 1755. This cooling occurs mainly because the gas is being made to do work, and this work is done at the expense of the kinetic energy of the molecules. If the gas could be allowed to expand into a vacuum, no cooling would occur in the case of an ideal gas, but a real gas, whose molecules exert attractive forces on each other, might be expected to cool, provided no heat was gained from the surroundings. The expansion of gases from a higher to a lower pressure through a jet or throttle was studied by J. P. Joule and W. Thomson (Lord Kelvin) in 1854. The fact that cooling occurs in many cases shows that real molecules exert an attractive force on each other. This force increases in magnitude as the molecules get closer together. There are several types of attractive forces known collectively as 'van der Waals' forces'.

Expts. 3a–2 and 4 show that a gas like sulphur dioxide may be liquefied in two ways: (1) by cooling, (2) by compression. Cooling reduces the dispersive effect produced by the thermal energy of the molecules; compression brings the molecules closer together, thus increasing the attractive forces between them; the gas condenses to a liquid when the van der Waals' forces overcome the thermal forces.

Sulphur dioxide was first liquefied by cooling (at atmospheric pressure) in 1800 (Monge and Clouet), and by compression in 1806 (Northmore). In 1823 Faraday, using both cooling and compression, liquefied a number of gases including chlorine, hydrogen chloride, hydrogen sulphide, ammonia, etc. The gas was produced under pressure from reagents placed in one limb of a sealed, bent tube, the other limb being immersed in a freezing-mixture. The method was rather dangerous and frequent explosions occurred, in one of which, thirteen fragments of glass entered Faraday's eye. Faraday failed to liquefy oxygen, hydrogen, nitrogen, carbon monoxide, etc., and these came to be known as 'permanent gases'.

The reason for this failure was made clear by Andrews's studies of carbon dioxide in 1869. He found that above 31·1 °C carbon dioxide could not be liquefied by pressure alone, but could be liquefied by pressure below this temperature. The existence of such a 'critical temperature' was first noticed in 1822 by Cagniard de la Tour for some volatile liquids such as ether. The critical temperatures (in deg C) for a number of gases are shown below:

Hydrogen	−239	Hydrogen chloride	−51·4
Nitrogen	−147	Hydrogen sulphide	100
Carbon monoxide	−139	Ammonia	131
Oxygen	−119	Chlorine	141
Nitrous oxide	−36·5		

It is now clear why Faraday failed to condense the so-called 'permanent gases'. Some of these were liquefied by Pictet in 1877 by cooling in liquid carbon dioxide boiling under reduced pressure, and also by Cailletet, about the same time, by allowing the compressed gas to expand suddenly and to do external work, thus losing energy and cooling.

The Joule-Thomson effect was applied to the liquefaction of air on the large scale by Linde in Germany and by Hampson in England (1894). With the liquefaction of hydrogen (at $-253°C$) by Dewar and by Ramsay and Travers in 1895, most gases had been condensed by the end of the nineteenth century. Helium, with its very low critical temperature, remained unliquefied until 1907, when it was liquefied by Kammerlingh Onnes at $-268·9°C$. It was solidified by Keesom in 1926 at $0·89°K$.

THE KINETIC THEORY AND THE LIQUID STATE

Expt. 3a-3

Fill a small bottle with a concentrated solution of potassium permanganate and place it, unstoppered, in a large dish or beaker. Run water very slowly into the latter so that the bottle is completely immersed. Leave undisturbed for some hours. The permanganate solution will slowly diffuse throughout the water.

Like the molecules of a gas, those of a liquid or a solute are in a state of rapid motion. This accounts for the free diffusion of a solute throughout its solvent as illustrated in the above experiment. The state of agitation of the molecules of a liquid is well brought out by the observation, first made by the botanist Brown in 1827, of the irregular movement of minute pollen grains in water due to the uneven bombardment of the grains by the water molecules. This phenomenon, known as the 'Brownian motion', may easily be observed with a number of suitable suspensions under a high-power microscope (see Expt. 3a-5).

The studies of Andrews (1869) showed a certain 'continuity of state' between the liquid and gaseous forms of a substance, and physical chemists of the next half century treated liquids as if they were condensed gases. However, the view that the molecules of a liquid are distributed completely at random, as in a gas, has undergone considerable modification since about 1930. At that time, X-ray studies of liquids began to show that the constituent particles were not distributed in so random a manner that no X-ray pattern is produced. The X-ray pattern for liquids is much more diffuse than for solids, indicating a far greater disorder among the particles of a liquid, but where X-ray pictures for both the liquid and the solid states of a given substance can be obtained, there is a general resemblance between them. The inference is that, when a solid melts, the crystalline pattern is not entirely lost, and that in a liquid some degree of order among the particles persists. As the liquid is cooled, a greater orderliness is introduced, until, near the freezing point, small aggregates of particles exist ready to fit together to form a crystal when the liquid solidifies.

According to this view, a molecule of a liquid is all the time within a field of force of about as many other molecules as in the solid phase; it does not, as in a gas, spend most of its life in comparative freedom and interact strongly with other molecules only occasionally.

Expt. 3a-4 Condensation of sulphur dioxide by pressure

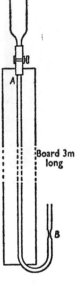

Displace the air from the tube by sulphur dioxide and seal off the capillary at *B* (Fig. 11). Pour mercury into the tube at *A* in such a way that the sulphur dioxide does not escape but is compressed. It will liquefy in the capillary when the height of the mercury column is about $2\frac{1}{2}$–3 m. A suitable internal diameter for the tube is about 3 mm.

Expt. 3a-5 Brownian motion

(1) Place on a microscope slide a minute speck of gamboge from an ordinary water-colour tube. Moisten with water, cover with a glass cover-slip and examine under a high-power microscope. As alternatives, a drop of diluted indian ink or colloidal graphite may be used.
(2) Add aqueous ammonia to a little weak silver nitrate solution in a test-tube. (The strengths of the weak solutions used are important.) Add a drop or

Fig. 11

two of a weak gelatine solution and a little weak potassium bromide solution. Dilute to a slight turbidity with distilled water. One drop of the resulting suspension is placed on a slide, covered with a cover-slip, and examined under a microscope using a high-power objective (4 mm). The agitation of the small particles can be very easily seen.

(b) Vapour pressure

Expt. 3b-1

Place 1 or 2 cm³ of ethyl ether on a watch-glass supported on a stand, and hold a lighted taper below and to one side of the watchglass. Note that the ether ignites when the flame is still some distance from the liquid.

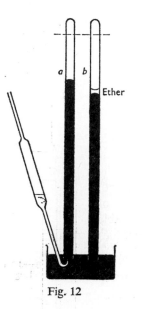

Fig. 12

Expt. 3b-2

Fill a wide barometer tube with mercury and invert it over a trough of mercury. By means of a bent capillary tube, introduce a drop of ethyl ether into the vacuum above the mercury in the tube. Note that the ether is all converted to vapour and measure the fall in the mercury level. Allow another drop of ether to enter the tube, and note a further fall in the mercury level. The space above the mercury contains ether vapour which is exerting a pressure equal to the observed fall in the height of the mercury, but the space is not saturated with vapour (Fig. 12, *a*). Continue to pass small quantities of ether into the tube until a small drop of liquid ether remains unevaporated on top of the mercury (Fig. 12, *b*). The total fall in the mercury level is then a measure of the saturation vapour pressure of the ether at the temperature of the liquid ether. Observe the changes in pressure as the outside of the tube near the liquid is warmed, or cooled. The cooling may be most easily effected by wiping the outside of the tube with a little ether.

The existence of vapour above the surface of a liquid shows that at least some of the molecules in the liquid are able to overcome the forces holding them together and to escape from the liquid into the space above it. If the kinetic theory is extended to the liquid state, the following picture of the phenomena associated with vapour pressure may be built up.

When a gas is condensed, the normal probability distribution of kinetic energy among the molecules of the gas persists in the liquid state, i.e. some molecules in the liquid will, at any given moment, have a small kinetic energy, many will have energies near an average value, and a few will have exceptionally high energies. The molecules of a liquid are in close proximity and their attractive forces hold them together in spite of their thermal motion. If molecules are to escape from the liquid and become gaseous, they must have sufficient kinetic energy to overcome these attractive forces. There are always some molecules having a sufficiently high kinetic energy to do this, and their number will increase with temperature. This is the kinetic picture of the process of evaporation and of the increase in rate of evaporation with temperature.

The fraction of molecules with sufficient energy to vaporize at temperature $T°K$ is given approximately by the Boltzmann distribution

$$n/n_0 = e^{-L/RT}$$

where L is the least energy required for 1 mole to vaporize at $T°K$. L is constant over small ranges of temperature, and approximately equal to the latent heat of vaporization per mole.

If the liquid is allowed to vaporize into a closed space, a concentration of vapour will build up. Molecules will evaporate at a rate depending on the temperature, and will recondense at a rate governed by the rate of bombardment of the liquid surface, i.e. at a rate proportional to the pressure of the vapour. When this rate of condensation becomes equal to the rate of vaporization, a state of equilibrium is set up, and a steady 'vapour pressure' is established, which is constant at constant temperature. Thus, when liquid and vapour are in equilibrium, molecules are continuously leaving the liquid and entering the vapour phase, and also leaving the vapour and entering the liquid, the rates of evaporation and of condensation being equal. For this reason, the state of equilibrium is termed 'dynamic'.

This conception results directly from the kinetic theory, and explains the readjustment of such equilibria when they are disturbed by alterations in pressure and temperature. The effect of such alterations is expressed by le Chatelier's principle, which states that: 'When a system in equilibrium is subjected to a stress, the equilibrium position shifts in such a direction as will tend to relieve the stress.' Thus, in the above example, increase in temperature shifts the equilibrium in the direction of the formation of more vapour, for in evaporating into vapour, the liquid absorbs heat, and vaporization thus helps to take up the 'strain' imposed by the raising of the temperature. A rise in temperature increases both the rate of evaporation and the rate of condensation, and results also in an increased vapour pressure.

The measurement of the vapour pressures of a liquid at different temperatures affords a means of measuring the value of L, the energy

E B C—E

of vaporization. The rate of vaporization, r, at any given temperature, is proportional to n, the number of molecules having energy greater than L. So r is proportional to $n_0 e^{-L/RT}$, and since r is proportional to p, the vapour pressure, p is proportional to $n_0 e^{-L/RT}$,

$$\log_e p = -L/RT + \text{a constant.}$$

If $\log_e p$ is plotted against $1/T$, the slope of the straight line will be $-L/2 \cdot 303R$, whence L may be calculated.

The following experiment describes how L may be found in this way, and compared with the latent heat of vaporization as measured directly.

Expt. 3b-3　Change of vapour pressure with temperature

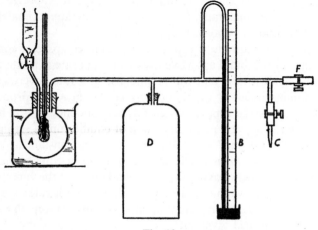

Fig. 13

The apparatus consists of a small bolt-head flask A (Fig. 13) fitted with a rubber bung carrying a thermometer (100 °C), an exit tube, and a tap-funnel with a gas-tight stopcock and a bent stem as shown. D is a large bottle of about 2 litres capacity, and C is a tap or screw-clip on a pressure-tubing connexion to a small glass jet. A little cotton wool is tied round the bulb of the thermometer. Some water, ethyl alcohol or other suitable liquid is placed in the tap-funnel and the flask surrounded by a water-bath.

Remove the air from the apparatus with a water-pump (or an oil-pump) until the pressure is about 2 cm of mercury. Close tap F. If the apparatus is air-tight, there will be no appreciable change in pressure in 5 min. Allow a little liquid to run on to the cotton wool, and heat the water-bath until its temperature is about 10 deg higher than that registered by the thermometer in A. When the latter is steady, the liquid alcohol around the thermometer is in equilibrium with the alcohol vapour in flask A. The pressure p, exerted by this vapour must be equal to the air pressure in the rest of the apparatus, which is measured by the difference in the heights of the barometer and the manometer B. Therefore read the thermometer and the manometer when the steady state has been reached. Then allow a little air to enter the apparatus through C, so that the manometer reading falls about 5 cm. The temperature registered by the thermometer in A will rise until equilibrium is again established. Read the temperature and pressure. Obtain a series of readings in this way, maintaining the temperature of the water-bath 5–10 deg higher than that registered by the thermometer in A.

In order to determine the energy of vaporization, L, plot $\log_{10}p$ against $1/T$, where T is the temperature in degrees absolute. Measure the slope of the graph, S. The latent heat of vaporization per mole is then given by $L = 2.303S \times R$, where $R = 2$ calories per mole (see p. 39).

Specimen result

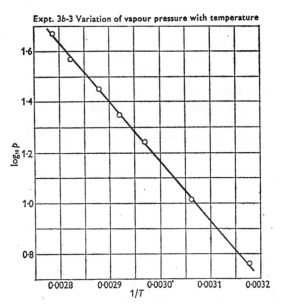

Expt. 3b-3 Variation of vapour pressure with temperature

Fig. 14

These measurements give a value of 10,600 calories per mole for the latent heat of steam over the temperature range 45–85 °C.

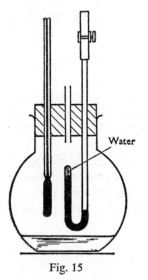

Fig. 15

BOILING

Expt. 3b-4

Set up the apparatus shown in Fig. 15 and gradually raise the temperature of the water in the bolt-head flask. Note that, as the boiling point is approached, the volume of vapour above the water in the J-tube increases owing to the increase of vapour pressure with temperature. At the boiling point of the water, the mercury levels in the two limbs of the J-tube should be equal.

The preceding experiment shows that the boiling point of a liquid is the temperature at which its saturation vapour pressure becomes equal to the external pressure on the surface of the liquid. Consider what happens when the temperature of the liquid in a beaker is steadily raised. Any small bubbles of gas formed at the surface of the glass under the water will be saturated with vapour at the saturation vapour pressure of the liquid. As the temperature rises, so the vapour pressure in the bubbles will rise. When this pressure exceeds the hydrostatic pressure in the surrounding liquid, the bubbles will push their way out. The deeper in the liquid that the bubbles form, the more violent will be the ebullition. If no air bubbles are present in the liquid, its temperature may rise well beyond its boiling point before any bubbles of vapour form; when the transition eventually occurs, the vapour formed will be at a pressure well above the external pressure and violent 'bumping' may take place. It is to provide bubbles of air and prevent this superheating that a piece of porous material is often placed in a liquid to obtain smooth and regular boiling.

Trouton's rule. For a number of liquids, the ratio of the latent heat of vaporization per mole at the boiling point, to the boiling point in degrees absolute has approximately the same value. The figures in the following table illustrate this relation, which is called 'Trouton's rule'.

Compound	Latent heat L at b.p. (Cal/mole)	T(K)	$\dfrac{L}{T}$
H_2SO_4	12·0	593	20·3
HCl	3·86	188	20·5
H_2O	9·82	373	26·3
NH_3	5·58	240	23·7
SO_2	5·96	263	22·6
CS_2	6·5	319	20·3
CCl_4	7·28	350	20·8
$CHCl_3$	7·0	334	20·9
$(CH_3)_2O$	5·14	248	20·5
CH_3OH	8·43	338	24·9
C_2H_5OH	9·22	352	26·2
C_6H_6	7·4	353	20·9
$SnCl_4$	8·3	387	21·4
$SnBr_4$	9·8	478	20·4
Fe_2Cl_6	12·0	588	20·4
Al_2Br_6	10·9	536	20·3

Note the constancy of the lower values of the ratio L/T at about 20·4. In view of the range of types of compound shown it is interesting to note that the molecules of those compounds for which the value differs from 20·4 are all polar to some degree (see p. 20). This polarity of the molecules causes an interaction between them, making the effective 'molecule' heavier and thus giving rise to a higher figure for L (because L is the product of the latent heat of vaporization per gram and the molecular weight).

Boiling point determination

The chief difficulty in determining the boiling point of a liquid, the avoidance of superheating, does not arise if the liquid includes air bubbles or bubbles of vapour. Landsberger's method (see Chap. 6), in which the liquid is raised to its boiling point by the latent heat of condensation from a stream of its own vapour, and the following two micro-methods, are simple examples of the application of this principle.

Expt. 3b-5 Siwoboloff's method

Prepare a tube (A, Fig. 16) about 3 or 4 mm internal diameter, 5 cm long and sealed at one end. Attach it with a rubber band to a thermometer reading in tenths of a degree. (A suitable band is

made by cutting off a 1 mm length from a piece of rubber tubing.)
Seal off one end of about 2 cm of capillary made by drawing out a test-tube, and drop it into *A*, open end downwards. Pour the liquid (e.g. benzene) into *A* until the capillary is nearly covered. Clamp the apparatus and immerse it in water in a large beaker to a depth of about 4 cm. Heat the beaker on a sand tray or electric heater and stir the water all the time so that the temperature rises steadily.

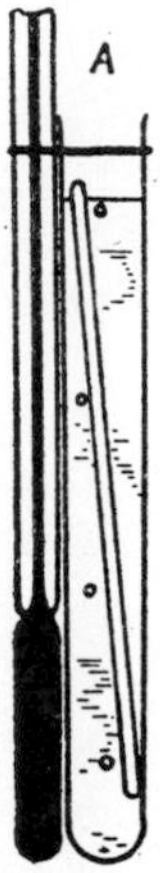

As the temperature rises, bubbles escape from the capillary tube, but when the boiling point is reached, a steady stream of bubbles emerges, and the temperature is then read. The boiling point may be more definitely determined by noting the temperature at which bubbles cease to emerge from the capillary as the liquid cools.

Fig. 16

(c) Liquid surfaces

Expt. 3c-1

Make a square frame of stout copper wire (see Fig. 17), and across opposite corners tie a piece of cotton with a loop in the middle. Dip the frame in a solution of soap or other surface-active substance (such as Teepol) and obtain a continuous film across the frame. Pierce the film within the cotton loop with a hot wire or pin. Note that the film contracts, pulling the loop of cotton into a circle.

Expt. 3c-2

Support a piece of iron or copper plate (15–30 cm square) on a tripod, and heat it strongly. Allow water to drop slowly on to the hot plate and note the spherical shape assumed by the drops.

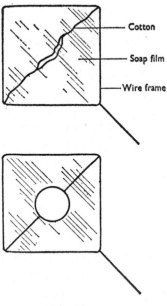

Cotton

Soap film

Wire frame

Fig. 17

Expt. 3c-3

Put about 50–100 cm³ of aniline into a litre beaker of water and slowly raise the temperature. As the temperature rises, the water becomes less dense, and at a certain temperature the aniline begins to sink. When this happens, remove the flame and gently stir the contents of the beaker so as to break up the aniline into large globules. Allow to stand and observe the behaviour and form of these drops of aniline.

The sphere has the smallest surface area of any solid form of given volume. Thus, these experiments show that liquid surfaces tend to contract to the smallest possible area. This property is accounted for very simply by the kinetic-molecular theory in terms of the attractive forces between the molecules. We have seen that the molecules of a liquid are like those of a gas in that they are free to move relative to one another, but like those of a solid in that they are kept close together by the forces of attraction between them. These forces are sufficient to prevent most of the molecules from escaping into the vapour phase. Within the body of the liquid, each molecule is surrounded by others and subject to forces of attraction in all directions, but at the surface of the liquid, the molecules are attracted to all sides and inwards, but this *inward* force is not opposed by an *outward* force.

Hence molecules in the surface of a liquid are subject to resultant forces directed inwards at right angles to the surface. This inward attraction results in a tendency for a liquid surface always to assume the smallest possible area, for molecules are continuously being pulled into the interior of the liquid. The obverse of this tendency is apparent in the fact that in order to increase the surface area of a liquid, work has to be done, i.e. additional energy must be supplied to form a new surface. There is thus a certain quantity of energy stored up in every unit area of surface.

In describing surface phenomena, it has been customary to say that surfaces behave as if consisting of an elastic skin, which tends to contract and which resists any attempt to increase the surface area. The term 'surface tension' is a century-old one and is used to denote a force acting in and parallel to the surface against which work must be done to increase the area of the surface. Thus if γ is the surface tension, the work done in joule in increasing the surface area by 1 m² is γ. However, this is a measure of the energy stored in 1 m² of surface. The concept of surface tension is thus a convenient means of measuring surface energy, but it is nothing more than a mathematical device. There is no 'elastic skin' at the surface of liquids, and no forces acting in and parallel to the surface; the behaviour which suggests their existence is due, as described above, to forces acting inwards and at right angles to the surface.

Sugden's method of comparing surface energies is described in Expt. 3c–4 below. It is a simplified version of Jaeger's method and is based on the relation between the excess pressure, p, inside a spherical bubble of gas in a liquid and the surface energy of the liquid, γ. This relation may be derived as follows.

The surface area of a spherical bubble of radius r is $4\pi r^2$. An increase in radius of dr is therefore accompanied by an increase in area of $8\pi r\ dr$. The increase in surface energy is $8\pi r\ dr\ \gamma$. The work done by the excess pressure inside the bubble when the latter expands is $4\pi r^2 p\ dr$. Since this work is absorbed as surface energy.

$$8\pi r\ dr\ \gamma = 4\pi r^2 p\ dr, \quad \text{i.e.} \quad p = 2\gamma/r.$$

Expt. 3c-4 Measurement of surface energy

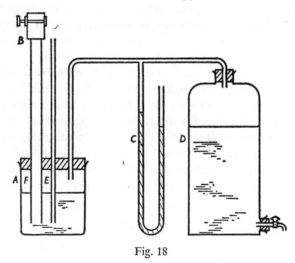

Fig. 18

The apparatus (Fig. 18) is a simplified form of Sugden's modification of Jaeger's method. Two tubes of different internal diameters, E and F (F being the larger), dip to the same depth in the liquid, of which only 4 or 5 cm³ are required. Bubbles can be blown on either tube by reducing the pressure in A. This is done by running water slowly out of an aspirator. The pressure is measured on the manometer C, using coloured water as the manometric liquid. With clip B open, bubbles will form on the larger tube (which should be between 3 and 4 mm internal diameter), and with the clip shut, bubbles form on the smaller tube (internal diameter about 0·2 mm).

To calibrate the apparatus, after thoroughly cleaning it with chromic acid followed by distilled water, place pure benzene in A, and form bubbles slowly, first on one tube and then on the other, noting the manometer readings, e and f. Use the formula $T = K(e - f)$, and, taking the surface tension of benzene to be 29×10^{-2} newton per metre, calculate the constant K.

Now determine the surface tension of (i) water and (ii) ethyl alcohol. Then investigate the effect of adding small quantities of 'surface-active' substances to the water. Try ethyl alcohol, amyl alcohol, phenol, fatty acids, soaps, Teepol solution in various concentrations from 0·05% by volume. The measurements are quickly made, and although the bubbles should be formed very slowly in theory, the results obtained in practice show that a determination can easily be done in 15 minutes. (There is no need to immerse the tubes to *exactly* the same depth each time.)

SURFACE FILMS

Expt. 3c-5

Completely fill a large clean rectangular dish or photographic tray, preferably black, with tap water. Put an oil film on the surface of the water by touching it with a glass rod dipped in old (oxidized) lubricating oil. Place a strip of glazed paper or a glass slide across one end of the dish in the surface of the water. By moving the slide steadily towards the centre, compress the oil film and note the changes in the interference colours of the film as its thickness is increased.

It has long been known that a small quantity of oil spread on the surface of water greatly lowers the surface tension of the water. Expt. 3c-6 below shows that not all kinds of oil will spread on water; thus pure paraffin oils do not, whereas oxidized lubricating oil and vegetable oils spread readily. The pure hydrocarbons do not mix with water at all and do not spread on the surface of water. (Neither does a drop of water spread on the surface of paraffin wax, whereas a drop of hexane does spread.) Substances whose molecules contain water-soluble groups, such as —COOH, —CH$_2$OH, etc. (long-chain acids and alcohols), will spread into surface films on water.

In 1890 Lord Rayleigh made the first quantitative measurement of the lowering of surface tension produced by a given quantity of oil. The following year Miss Pockels discovered that the surface tension of the water is only lowered if the oil is confined in a certain minimum area, and that if a given quantity of oil is allowed to spread over an area larger than this critical minimum area, the surface tension of the water is unaltered. In 1899 Lord Rayleigh suggested that under those conditions where the oil film is confined in the greatest area where it begins to affect the surface tension, the oil molecules form a layer just one molecule thick. This suggestion has been fully confirmed by the measurements of I. Langmuir in America and N. K. Adam in this country. The surface tension is a measure of the tendency to oppose extension of the surface. This is reduced by the presence of an oil film because the repulsion between the oil molecules opposes compression of the surface. As an oil film on water is compressed, the point at which the oil molecules first repel one another will be that at which the film is just one molecule thick. Because of this fact, the study of surface films has proved a simple way of studying the molecules themselves. Thus, during the 20 years following 1918, Langmuir, Adam, Rideal and others have studied surface films of many different long-chain molecules on water and on other substrates. From their work it is clear that the molecules are anchored to the water by their polar groups. In very 'dilute' films, the hydrocarbon chains are lying down in the water and the molecules are free to move about in the film like the molecules of a gas. If the film is compressed, the molecules 'stand up' with their polar heads still in the water but with their hydrocarbon tails in the air. Further compression locks the hydrocarbon chains into a two-dimensional solid. Considerable information about the dimensions of the molecules and the forces between them has been obtained. Further work has been directed to the study of the effect of the nature of the substrate and to chemical reactions occurring in surface films.

Expt. 3c-6

(1) Thoroughly clean a flat dish, preferably black, such as is used for photographic purposes. Fill the dish with tap water and scatter the surface evenly with powdered talc or flowers of sulphur from a

muslin bag. Put a drop of oleic acid or camphorated oil on the surface and note that it spreads out, pushing the powder aside.

(2) Put clean water in the tray and float a strip of paper on the surface to divide it roughly in half. Put a drop of oleic acid at one end of the tray and note that the paper barrier is repelled to the other.

(3) (*Caution!*) Make a saturated solution of ether in water and place it in the tray. Ignite it. Drop a little oleic acid on to the burning liquid. The oleic acid film will spread over the surface and extinguish the flame.

Expt. 3c-7   Estimation of the size of a molecule

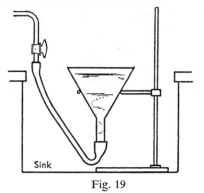

Fig. 19

The apparatus illustrated in Fig. 19 provides a simple means of obtaining and renewing a clean water surface.

(1) Cover the surface with powdered talc from a muslin bag and, with a glass tube, drop on a little medicinal paraffin dissolved in petroleum ether. The ether quickly evaporates, leaving a drop of the oil. Note that it does not spread. Flood the funnel liberally to produce a clean surface. Again dust with talc and add a drop of petroleum-ether solution of stearic acid or oleic acid containing 0·05 g/litre of 40°/60° petroleum ether. Note that spreading occurs. Measure the area of the surface film formed from one drop of the solution of stearic acid. Obtain the volume of one drop by allowing 1 cm^3 to drop from a burette and counting the number of drops. Assuming that the film is one molecule thick, make a rough estimate of the cross-sectional area of a stearic acid molecule. The number of molecules in 1 mole $= 6 \times 10^{23}$; molecular weight of stearic acid $= 284$. The area per molecule is about 20·5Å^2 (1 square Ångström unit $= 10^{-20}$m^2).

(2) An alternative way of performing this experiment is to float a loop of cotton on the water and to count the number of drops of stearic acid solution needed to fill the loop with a 'solid film' of fatty acid. Use the solution described above in (1). Tie together the ends of a piece of cotton about 30 cm long to make a loop and allow the loop to fall on to the surface of the water. Make sure that the cotton loop lies entirely on the surface with no gaps or parts of the cotton submerged. Use a teat pipette to drop the fatty acid solution on to the surface of the water within the loop. After 2 or 3 drops have been added, push gently against the loop with a glass rod. The cotton loop will be dented, but after the addition, one by one, of a few more drops, it will be found that the loop can be moved across the water surface as a solid raft. It can be assumed that the molecules of stearic acid now form a compact monolayer on the surface of the water within the loop. Note the number of drops added and calculate the cross-sectional area of a molecule as above.

4 The solid state

(a) Equilibrium between the liquid and solid states

Expt. 4a-1   Determination of freezing points by the methods of cooling

Melt enough of the solid (naphthalene is suitable) in a small test-tube to cover the bulb of the thermometer (100°C), and mount the tube in a small bottle or flask to protect it from draughts and ensure a uniform rate of cooling (see Fig. 20). Put a little cotton wool round the thermometer in the mouth of the tube and note the

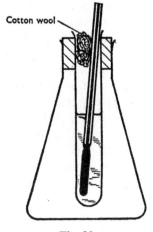

Cotton wool

Fig. 20

temperature every half-minute. Plot a graph of temperature against time. The temperature falls smoothly until crystals begin to appear. If supercooling has occurred, the temperature rises slightly and becomes steady until all the liquid has frozen. This steady temperature is the freezing point. (See Fig. 21.)

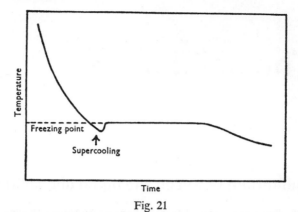

Fig. 21

When a liquid is cooled, its molecules lose heat energy and their movements become progressively less vigorous until the attractive forces between the atoms become predominant over the dispersive forces arising from their thermal motion and, at the freezing point, the atoms take up positions in a solid structure. Even in the solid state, the atoms, although arranged in a definite pattern, are not stationary; they still possess energy and vibrate about mean positions. This vibrational energy diminishes with fall in temperature, but only vanishes at the absolute zero. As the temperature is then raised, the vibrational energy and the amplitude of vibration increase. Thus, for sodium chloride, the amplitude just below the melting point is about $\frac{11}{10}$ths of the amplitude at ordinary temperatures. A further rise in temperature causes the atoms to break out of the crystal structure and the solid to melt.

At the melting point, solid and liquid are in dynamic equilibrium. We have had an example of the latter in the equilibrium between a liquid and its vapour (see Chap. 3b). The state of dynamic equilibrium that exists between a substance in the liquid and solid forms is clearly brought out by the determination of a freezing point by the method of cooling (Expt. 4a-1). As the hot liquid loses heat, its temperature, of course, falls until the liquid begins to crystallize. The temperature then remains steady until all the liquid has become solid. This is because the latent heat set free as the liquid solidifies makes up for the heat lost to the surroundings by convection and radiation. The rate of crystallization is in fact governed by the rate of loss of heat to the surroundings.

It is possible for a liquid to cool below its freezing point without crystallizing. Such 'under-cooling' may occur to the extent of a fraction of a degree, or, as in the case of ordinary glass, to the extent of several hundred degrees. (The term 'supercooled' is more usual, though less suitable, than 'under-cooled'.) Spontaneous crystallization of a glass sometimes occurs after the lapse of many years and is known as 'devitrification'.

Glass occasionally crystallizes during its manufacture owing to some fault in its treatment: minute crystallites may form, giving the glass a slight opalescence, or the material may become quite opaque as a result of heavy crystallization.

Expt. 4a-2 Determination of melting points

Prepare a capillary tube about 1 mm internal diameter and 5 cm long by drawing out a test-tube in a Bunsen or blowpipe flame. Seal off one end and, when it is cool, shake into it a little of the powdered solid whose melting point is to be determined (naphthalene or *m*-dinitrobenzene is suitable). Fix the capillary to the thermometer with a rubber band and support it in a 250 cm³ beaker of water. Raise the temperature of the water slowly, keeping it well stirred. When the melting point is neared, the rate of rise of temperature should be about 1 deg C per min. Note the temperature at which the crystals first begin to melt. Repeat the determination, making sure that the lowest temperature at which the crystals will melt is obtained. For a pure substance, the melting point is identical with the freezing point; but for mixtures, this is not so (see Chap. 5d).

(b) The crystalline state

Expt. 4b-1

Compare the physical forms of the following substances in their amorphous and crystalline states: (*a*) sulphur (amorphous and rhombic), (*b*) silica (silica gel or precipitated silica, and quartz crystals), (*c*) calcium carbonate (precipitated chalk and Iceland spar).

The above are just a few examples of substances that can exist in more than one solid form. In the crystalline state, the substance may exist in well-defined, geometrical forms and may cleave along certain planes at characteristic angles to one another. For example, a piece of crystalline calcite, if tapped gently with a hammer, will split up into small rhombohedra. A given substance can often be obtained in more than one crystalline form. In the amorphous state, the substance exhibits no planes of cleavage and no crystalline shape; for example, a piece of chalk is not bounded by flat faces and, when hit by a hammer, breaks irregularly.

The regular shapes of crystals are an outward expression of the regular arrangement of their constituent particles. In the amorphous state, the arrangement of the particles is random, but most amorphous substances are shown by X-ray analysis to consist of random mixtures of microcrystalline aggregates.

The work of the early crystallographers on the external geometry of crystals showed that all the many varied shapes and forms assumed by crystals could be conveniently classified into six systems according to the degree of symmetry. This may be measured by the number of planes of symmetry the crystal possesses, i.e. the number of imaginary planes which may be drawn through the crystal to divide it into two halves that are mirror images of each other. It is usual now to use a classification that includes a seventh system, namely, the trigonal or rhombohedral system, which was previously treated as a section of the hexagonal system in which only every alternate face of a given form is developed.

The most symmetrical forms, such as the cube or regular octahedron, have nine planes of symmetry and belong to the first, or cubic, system. Some characteristics of the crystal systems are shown in the table and in Fig. 22.

System	Max. no. of planes of symmetry	Crystallographic axes
Cubic	9	Three equal axes at right angles
Hexagonal	7	Three equal, coplanar axes at 60°, a fourth axis at right angles to the other three
Tetragonal	5	Three axes at right angles, two equal
Orthorhombic	3	Three unequal axes at right angles
Monoclinic	1	Three unequal axes, two at right angles, the third oblique
Triclinic	0	Three unequal axes, no two at right angles
Trigonal or Rhombo- hedral	3	Three equal axes, equally inclined but not at right angles

Each system includes a number of 'forms', for example, the cubic system includes the cube, the regular octahedron, the rhombic dodecahedron, etc., all of which have the highest degree of symmetry. A substance that belongs to the cubic system may crystallize in any of these

a

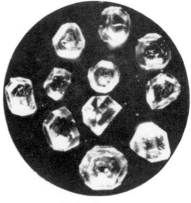

b

c

d

1   CRYSTALS

a Sodium chloride (cubic system) ($\times 10$)
b Sodium chloride grown from a solution containing urea: cubic crystals with
     octahedral facets ($\times 15$)
c Sulphur (orthorhombic system) ($\times 10$)
d Copper sulphate (triclinic system) ($\times 10$)

e

f

g

1 CRYSTALS (*contd*)

e **Beryl (hexagonal system). Synthetic emeralds** ($\times 5$)

f **Quartz (rhombohedral system)** ($\times 10$)

g **Calcite (rhombohedral system)** ($\times 10$). Three of the crystals rest on a hair. A
    double image can be seen through two of them, the optic axis of the third
    crystal is roughly parallel with the hair and therefore only one image is visible

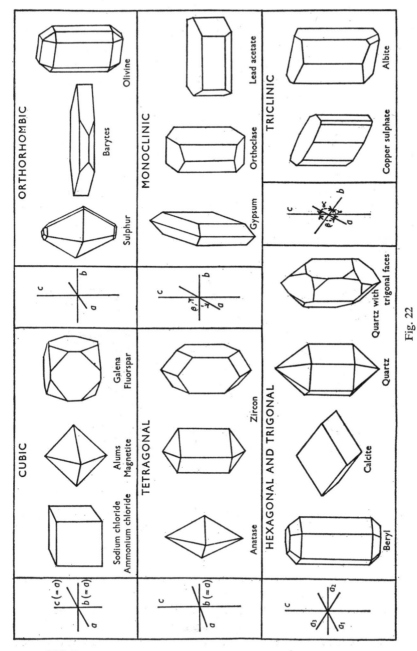

Fig. 22

forms or in combinations of them. Thus, the minerals galena and fluorspar occur in cubic crystals and in crystals consisting of combinations of cubes and octahedra. Common salt may also be crystallized in both cubic and octahedral forms and in combinations of the two forms (see Expt. 4b-3). It is exceptional to find perfect forms, either in nature or in crystals grown in the laboratory. Some faces grow faster than others, and growth seldom takes place uniformly for a variety of other reasons.

The faces of a crystal meet at definite angles, and it has been known for many years (Nicolas Steno, 1669) that, for a given substance, the angles between the faces of a particular form are constant in all crystals of the substance, however irregularly the crystal may have grown.

The position of any crystal face is defined by reference to the crystal axes. These differ from one system to another; particulars are shown in the table. For further details on the geometry of crystals, reference should be made to books on crystallography or mineralogy.

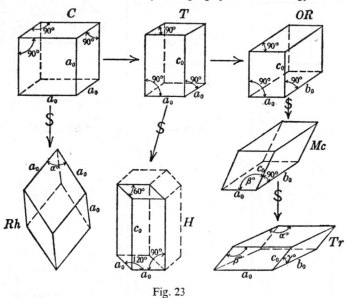

Fig. 23

The concept of the 'unit cell' was introduced by the Abbé Haüy (1784), who is usually regarded as the 'father of crystallography'. A unit cell has been defined as 'the smallest molecular unit by the multiplication and juxtaposition of which the homogeneous crystalline structure may be indefinitely extended'. Fig. 23 shows how the various types of unit cell may be regarded as interrelated by appropriate modifications of edge lengths or by 'skewing' (indicated in the figure by the sign $). In the relationship between the tetragonal and the hexagonal unit cells, the twist is into a special position characterized by the fixed angle 120°, whereas in the other cases the twist is into general positions.

Certain substances crystallize in more than one crystalline system. Sulphur is a familiar example, forming rhombic crystals below 96°C and monoclinic crystals above that temperature. This phenomenon is known as polymorphism and is treated in Chapter 4c.

THE STRUCTURE OF CRYSTALS

Our knowledge of the arrangements of atoms in crystals comes through the invention by W. H. and W. L. Bragg of the X-ray spectrometer, which makes use of the fact that a beam of X-rays is diffracted by the regular layers of atoms or ions in a crystal to give a diffraction pattern.

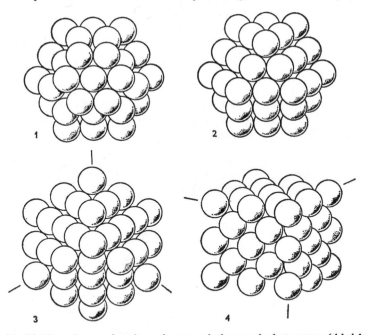

Fig. 24. (1) Three layers, forming a hexagonal close-packed structure (third layer directly above the first a,b,a). (2) There is an alternative position for the topmost layer, giving a different structure (a,b,c). (3) The same structure as (2), with more atoms added. (4) Another view of (3), to bring out the cubic character of the structure.

This can be used to determine the structure of the crystals. For more detail, reference should be made to books on crystallography.

The shape of the crystal is determined by the arrangement of its constituent particles. These may be atoms, ions or molecules. The most familiar examples of crystals composed of arrays of atoms are those of the metals. In crystals of a pure metal all the atoms are the same size and take up a very simple, close-packed arrangement. Two such arrangements are illustrated in Fig. 24. One is obtained by stacking 'rafts' of spheres *a-b-a* and is called 'hexagonal' close-packing, and the other by packing them *a-b-c*, and is called 'cubic' close-packing, or face-centred cubic structure.

Fig. 25. Body-centred cubic structure. Crystals of sodium, iron, and tungsten have this structure.

Another arrangement, known as the body-centred cubic structure, is shown in Fig. 25. A more open arrangement of atoms is found in, for example, diamond, where the structure is as shown in Fig. 26.

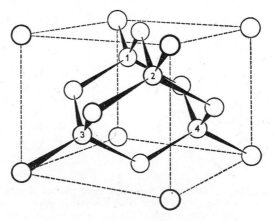

Fig. 26. The crystal structure of diamond.

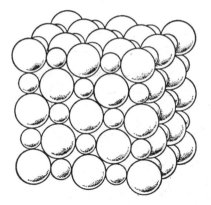

Fig. 27. Structure of a crystal of sodium chloride. The smaller balls are positively charged sodium ions, the larger ones negatively charged chlorine ions.

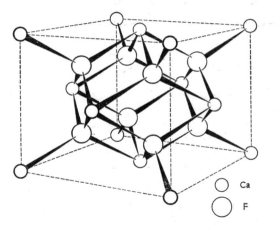

Fig. 28. The fluorite (CaF$_2$) structure.

The structure of ionic crystals is determined by the necessity for electroneutrality and by the relative sizes of the ions. Thus the sodium ion has a diameter about half that of the chlorine ion, and the packing of the ions is such that each has six 'nearest neighbours' as shown in Fig. 27. In the case of calcium fluoride, there must be two fluorine ions for every calcium ion, and it is clear from Fig. 28 that each calcium ion has four 'nearest neighbours'. In both examples, the external symmetry is cubic. The arrangement of the ions is extended throughout the crystal and is known as a 'giant structure'. Giant structures are often considerably more complex and the silicate minerals may be quoted as examples.

Thus mica has a structure in which double layers of a silicon–aluminium–oxygen framework are joined by layers of potassium and hydroxyl ions, as illustrated in the diagram (Fig. 29).

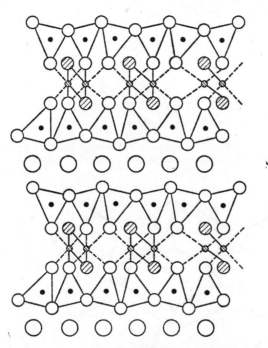

Fig. 29. The structure of mica.

i. Complex layers of silicon (Si), aluminium (Al), oxygen (O) and hydroxyl (OH). Separated by large ions of potassium (K$^+$).
ii. Complex layer. (a) A row of Si-O tetrahedra, slightly kinked (4th O not shown).
               (b) An OH after every 2 apexes of the tetrahedra.
               (c) An Al below or above every OH.
               (d) Al joined by thin continuous line to 2 (OH) and by thin dotted line to 4 (O) belonging to the tetrahedra.

Thirdly, the crystal may be composed of molecules. Crystals of many organic compounds e.g. naphthalene (Fig. 30) are of this kind. The molecules are held together by relatively weak forces and such crystals have low melting points compared with those of metals and of crystals consisting of giant structures. Another example, sulphur, is shown in Fig. 31.

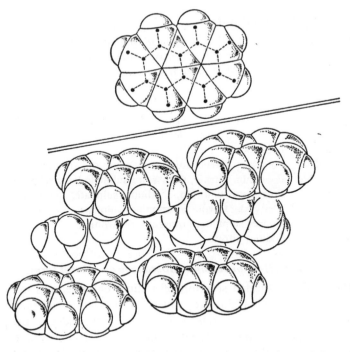

Fig. 30. *Above:* a molecule of naphthalene; *below:* the arrangement of molecules in the crystal.

Fig. 31. Arrangement of molecules in a crystal of sulphur. (Only a part of the complete arrangement is shown.)

Expt. 4b-2   The crystal systems

Prepare warm saturated solutions of the following salts and place a drop of each on a clean microscope slide. Leave to crystallize and examine under a hand lens or low-power microscope.

System	Substances
Cubic	Sodium chloride, potash alum, potassium cadmicyanide, sodium chlorate, lead nitrate
Hexagonal	Lead iodide
Tetragonal	Potassium ferrocyanide, potassium dihydrogen phosphate
Orthorhombic	Potassium nitrate, potassium sulphate, zinc sulphate
Monoclinic	Potassium chlorate, sodium sulphate, potassium ferri-cyanide, ferrous ammonium sulphate
Triclinic	Copper sulphate, potassium dichromate
Trigonal	Sodium nitrate

Expt. 4b-3   The effect of foreign substances on crystal form

(1) Make a saturated solution of sodium chloride, and in 10 cm³ dissolve 2 g of urea. Put drops of the two solutions on microscope slides and leave to crystallize. Examine them under a low-power microscope. The pure salt solution deposits rectangular crystals, whereas from the other solution, the salt separates as cubes and octahedra and combinations of these forms (see Plate 1). Try the effect on the crystal form of a trace of potassium ferrocyanide.
(2) Make a saturated solution of ammonium chloride and to 10 cm³ add 0·5 g of urea. Put drops of the two solutions on microscope slides and leave to crystallize. The salt separates in fern-like forms from the unadulterated solution and in regular cubes from the other.
(3) Dissolve about 14 g of magnesium sulphate crystals in 20 cm³ of distilled water and divide into two parts. To one, add 1 cm³ of a 10% solution of borax. Place the solutions in beakers, put filter-papers over them to exclude dust, and set aside to evaporate slowly. Alternatively, place them in a desiccator containing sulphuric acid. When crystals have formed, remove some and examine them with a low-power microscope. Those from the solution containing borax will be long four-sided prisms, with two terminal faces at each end, whereas in those from the other solution, the latter faces predominate to the virtual suppression of the prism faces.

CRYSTALLIZATION

Expt. 4b-4

(*a*) Melt a little *m*-dinitrobenzene on a microscope slide, cover with a warm cover-slip and allow to cool until crystals are obtained. Examine them with a hand lens.

(*b*) Warm a drop of a solution of this substance in benzene on a glass slide, cover with a warm cover-slip and leave to crystallize. Compare the crystals with those formed in (*a*).

Expt. 4b-5 Supercooling

Prepare two glass tubes, closed at one end, each about 30 cm long and 2 cm in diameter. Clean them with dichromate and sulphuric acid and wash them with distilled water. Fill the tubes with sodium thiosulphate crystals and warm by immersion in hot water until the crystals dissolve in their own water of crystallization. Set aside to cool with plugs of cotton wool in the open ends to exclude dust. When the tubes are quite cold, drop a crystal of sodium thiosulphate into one and a crystal of potassium nitrate into the other. The 'seed' of thiosulphate initiates the growth of crystals which spread down the tube in beautiful dendritic forms. The crystal of potassium nitrate should be without effect. Why does the former tube get hot?

A substance may be obtained in the crystalline state either by solidification from the liquid state or by evaporation of the solvent from a saturated solution. In general, the crystalline product is a mass of interlocking crystalline forms which are not usually recognizable as separate entities. In order to obtain separate crystals of true form, small pieces of crystal are 'grown' by immersing them in a saturated solution and allowing the solvent to evaporate regularly and slowly. The solid then deposits on the crystal nucleus and builds upon it a larger crystal with facets that conform recognizably to forms in the system to which the substance belongs.

When crystals grow either from a melt or from solution, the substance deposits on solid nuclei present in the liquid. If there are no nuclei, a considerable degree of supercooling may occur before solidification begins. Once some solid nuclei form (or if some are introduced), the crystals grow from these centres, sometimes uniformly in all directions forming granules, sometimes more rapidly in one direction than another forming plates or needles, and often in branching, fern-like shapes known as dendrites (see Plate 2). If there is no second substance present, the grains, dendrites, etc., grow until they meet their neighbours and the whole mass is crystalline. The material that has grown around one nucleus is called a crystal 'grain', and these meet at the so-called grain boundaries. Look for these features in (*a*) the crystallization of organic substances prepared in the laboratory; (*b*) crystalline aggregates of natural minerals; (*c*) metallic crystals, e.g. zinc on galvanized iron.

The structure of the crystalline mass will vary with the number of nuclei in the crystallizing liquid. If there are only a few nuclei, the size of the crystal grains will be large, whereas if crystallization starts at many nuclei, the grain size will be small. Rapid cooling favours supercooling, and hence the formation of a large number of nuclei with consequent small grain size. Slow cooling, and the introduction of a few nuclei from outside ('seeding'), favours growth from a few centres only, with the production of larger crystals. Thus by rapidly cooling a saturated salt solution, a finely crystalline product is obtained, whereas the crystals are much larger if allowed to form slowly.

Again, in the solidification of metals and alloys from a melt the rate of cooling affects the grain size, which in turn is an important factor in determining the physical characteristics of the solid alloy. For example, with a pure metal, slow cooling and annealing lead to large grain size with the production of a soft, malleable solid, whereas rapid cooling or 'quenching' leads to smaller grain size and a harder, more brittle solid of greater strength. For further details, see books on metallography.

THEORIES OF CRYSTAL GROWTH

Since solid nuclei within the solution will be uniformly bombarded by solute particles, one might expect that solid would be added to the nuclei in a regular manner and that this would lead to the formation of spheres. Under certain conditions of crystallization this does in fact occur and examples of spherulites are shown in Plate 3.

Although crystals with flat faces are more familiar than spherulites, their development is harder to explain. As a crystal grows on a nucleus, a corner or protuberance may develop, which will grow more rapidly because it is reaching out into regions where the supersaturation is higher, the solution being less exhausted by crystal formation there than in the immediate neighbourhood of the surface. This leads to the formation of branching forms or 'dendrites', a common and familiar phenomenon in the growth of crystals. However, the fact that dendrite formation does not invariably occur, but that flat faces can, under certain conditions, continue to grow shows that some mechanism other than direct deposition of solute on to the seed must be operating. Experimental study of growing crystals has shown that the newly-adsorbed particles can migrate over the crystal surface before being incorporated into the crystal structure. Thus, material arriving at, say, an edge may not stay there but may move within a mobile surface layer over the face of the crystal until it finds a favourable situation such as a step or ledge, at which to become part of the solid crystal. This movement of adsorbed material over the crystal surface is an important factor in theories of crystal growth.

The manner in which flat faces grow has been greatly illumined by the work of Bunn and Emmett who studied crystals growing in solution on microscopic slides. The growing crystals were observed with a polarizing microscope between crossed Nicols. These studies showed that the crystals grow by the addition of layers to the faces, these layers starting at some point on the crystal surface and spreading until they cover the whole face. New layers often start before their predecessors have reached the edges of the face, thus a series of steps, or a 'growth pyramid', forms on the growing surface. The substances studied by Bunn included sodium chloride, cadmium iodide, and ammonium dihydrogen phosphate. The surfaces of crystals of minerals often show these growth pyramids (see Plate 4).

Expt. 4b-6

Place a drop of a solution of ammonium or potassium dihydrogen phosphate on a microscope slide. Warm it gently and lower a coverslip onto it. Place it on the stage of a polarizing microscope and observe it between crossed polaroids, using a 6 mm objective. Move the slide until a growing crystal is in the field of vision. Look for changes in the polarization colour of the crystal, indicating changes in its thickness which occur as it grows. Can you see any places from which growth appears to start? Can you see evidence of growth sheets spreading across the crystal?

Theories as to how crystals with well-developed plane faces continue to grow have been put forward for years. Bunn and Emmett's photographs strongly suggest that a nucleus forms on the crystal face from which a sheet spreads across the face until it reaches the edge, thus adding another layer of uniform thickness to the crystal. The theory of growth by the formation of monomolecular surface nuclei, which then spread across the face, was first put forward by Gibbs in 1878. Burton and Cabrera (1949) calculated that the rates of growth would not become appreciable in supersaturations of less than about 50%. This is far from being in agreement with experiment, for the best crystals grow in supersaturations of 1% or even less. In other words, the rates of growth observed in practice are very much greater than those predicted by these theories.

In the same year, 1949, F. C. Frank outlined a theory which was a development of those mentioned above. Instead of assuming an 'ideal' structure for the crystal, he made use of the imperfections which are known from other evidence to exist in real crystals. These often have a mosaic structure, and contain 'dislocations' arising from discontinuities in the regular rows of atoms. One type of dislocation, a 'screw dislocation', gives rise, if it ends on a crystal face, to a ledge on the face starting at the point where the dislocation emerges and terminating at the edge

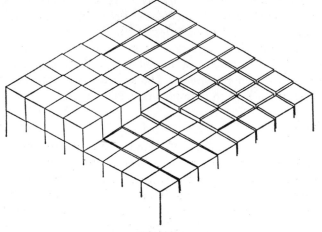

Fig. 32

of the face (see Fig. 32). Growth can occur by addition of material at the ledge. Surface nucleation is then unnecessary and crystals can grow in quite low supersaturations. Frank has discussed the forms the ledge will take as growth occurs. As material is added to it, the ledge will advance, but since it ends at the point where the dislocation emerges at the crystal face, the rate of growth at this end of the ledge will be less than else-where. Thus when growth is occurring, the ledge can only advance by rotating round the dislocation point. A spiral-shaped ledge thus develops on the crystal face. When two dislocations of opposite 'hand' occur near each other on the same face, the two spirals soon join and form a closed ring. The ring will then retain its shape as it spreads across the face of the crystal (see Fig. 33).

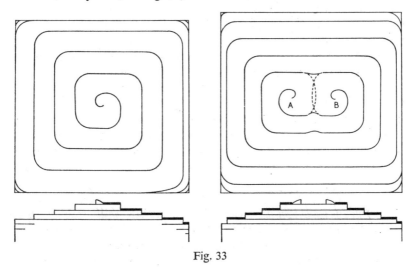

Fig. 33

At the time this theory was formulated no experimental evidence for spiral growth marks on crystals was known, but L. J. Griffin had already taken photographs of the surfaces of beryl crystals showing growth marks of the precise form indicated by Frank (see Plate 5a). This dramatic evidence for the screw dislocation theory of crystal growth was presented at the Bristol University Summer School in 1949 and published in 1950. Since then many spiral growth marks have been observed on both natural mineral crystals and on crystals grown from solution in the laboratory (see Plate 5b).

ISOMORPHISM

Expt. 4b-7

Prepare crystals of potash alum and chrome alum by allowing strong solutions to evaporate at room temperature. In making the chrome alum solution, do not heat it above 60°C.

Both substances crystallize in the cubic system and can be obtained in the form of octahedra.

Substances that crystallize in the same system are said to be 'isomorphous'. Isomorphous substances will (*a*) form 'overgrowths' (Expt. 4b-8), (*b*) form mixed crystals (Expt. 4b-9), and (*c*) act as 'seeds' for relieving the supersaturation of solutions of each other. Mixed crystals are better called 'solid solutions'. Many examples of substances that are completely or partially miscible in the solid state are to be found among the metals (e.g. copper and zinc in brass) and the naturally occurring minerals (e.g. sodium aluminium silicate and calcium aluminium silicate in the plagioclase felspars) (see Chap. 5d).

Expt. 4b-8   Overgrowths

Select a good crystal of chrome alum and place it in a saturated solution of potash alum in a desiccator. The potash alum will build on to the chrome alum, forming a larger crystal whose faces are parallel with those of the original octahedron.

Expt. 4b-9   Solid solutions

Make a mixed solution containing potash alum and a small quantity of chrome alum and leave it to crystallize. The resulting crystals will be a pale mauve colour intermediate between the colour of the two alums, the chromium and aluminium atoms being incorporated in one and the same crystal structure.

Expt. 4b-10   Oriented overgrowths

Place a small cleaved rhomb of calcite on a microscope slide with a freshly cleaved face uppermost. Etch the face of the calcite with a drop of acid and wash with distilled water. Cover the crystal with saturated sodium nitrate solution, and set aside to crystallize, protected from dust. Examine the crystals with a lens or low-power microscope. Most of the sodium nitrate crystals will be alined along the cleavage directions of the calcite.

This is an interesting example of isomorphism because the two substances are so unlike chemically. In the case of potassium sulphate and chromate, or potash and chrome alum, one atom replaces the other similar atom (S and Cr), (Al and Cr) in the crystal. However, in spite of the chemical dissimilarity of calcium carbonate and sodium nitrate, the sizes and arrangement in the crystal of the calcium and sodium ions and of the carbonate and nitrate ions are very similar.

(c) Polymorphism

Expt. 4c-1

Spread some mercury(II) iodide on a filter-paper and warm it gently above a hot-plate or sand-tray. The colour changes from red to yellow. When it cools, the substance remains yellow for some time, but turns red again on being touched with a glass rod.

Many substances exist in more than one crystalline form. Mercury(II) iodide, sulphur, carbon, phosphorus and tin will be familiar examples. In the case of mercury(II) iodide, the red form changes into the yellow at a certain temperature known as the transition point. On cooling the yellow form, however, it does not revert to the red at the transition point, but may be cooled to room temperature without change. It is then said to be in a 'metastable' state, and will slowly change into the red form on standing. The transition is facilitated by mechanically disturbing the metastable substance.

Transition points may be determined by making use of changes in some physical property (such as colour or density) which occur as the solid passes from one crystalline form to the other. The measurement of transition points by two methods is described in the following experiments.

Expt. 4c-2   Colorimetric method for copper(I) mercury(II) iodide

The temperature at which the red form of this compound changes into the black form may be observed by heating some of the substance in a melting point tube in a water-bath. The colour change is very striking. The substance is prepared from mercury(II) iodide and copper sulphate by reduction with sulphur dioxide. Dissolve 6·8 g of mercury(II) chloride in 300 cm³ of hot water and add a solution of 8·3 g of potassium iodide in about 50 cm³ of water. Allow the precipitate to settle, wash once by decantation, and dissolve it in a solution containing 8·3 g of potassium iodide in 50 cm³ of water. Add a solution of 12 g of copper sulphate in 150 cm³ of water and pass sulphur dioxide until no more is absorbed. Filter and wash the red precipitate and dry it in the steam oven.

To determine the transition point, place some of the dry powder in a melting point tube, attach it to a thermometer and warm it slowly in a well-stirred beaker of water. Note the temperature when the colour changes. The change back on cooling is not so sharp because some supercooling occurs.

Make some silver mercury(II) iodide and examine the action of heat on it in a similar way.

Expt. 4c-3   Dilatometric method for sulphur

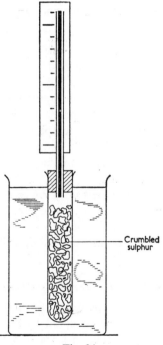

Crumbled
sulphur

Fig. 34

This method depends on the fact that there is an abrupt change in volume at the transition point of rhombic to monoclinic sulphur.

Carefully heat about 5 cm³ of 30% sulphuric acid in order to free it from dissolved air. Allow it to cool in a stoppered bottle. Fill about a quarter of a strong test-tube with the acid and drop in coarsely powdered roll sulphur until the test-tube is nearly full. Place it in a beaker containing a 30% solution of calcium chloride and warm it to about 80°C, allowing the acid to overflow into the

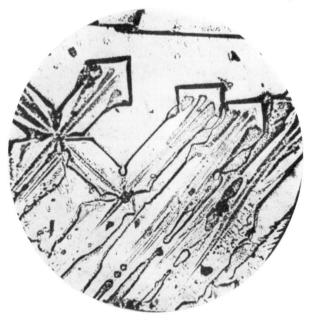

a

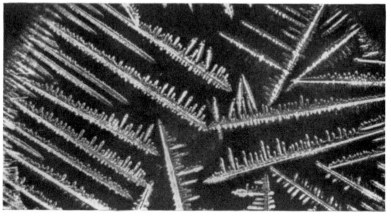

b

2 DENDRITIC CRYSTALLINE GROWTH

a Dendritic crystals of lead nitrate showing cubic terminations ($\times 50$)
b Ammonium chloride (cubic system) growing from aqueous solution ($\times 25$)

a

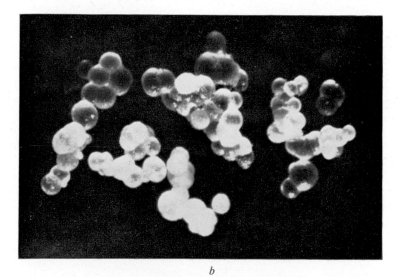

b

3 RADIAL CRYSTALLINE GROWTH

a *Top left:* Kidney iron ore (haematite)   *Top right:* Pyrites nodule (split open)
    Bottom left: Pyrites nodule   *Bottom right:* Partially devitrified
      (all ×½)   glass
b Synthetic critobalite (×50)

a

b

c

4   GROWTH SHEETS ON THE SURFACE OF MINERALS

a   Topaz
b   Prism face of natural quartz (×40)
c   Rhombohedral face of natural quartz (×40)

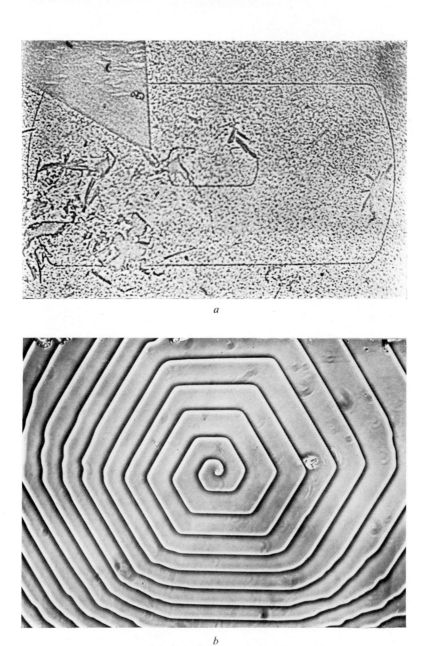

a

b

5 SPIRAL GROWTH SHEETS

a   Prism surface of beryl (Griffin) ($\times 700$)
b   Spiral on a SiC crystal (Verma) ($\times 130$)

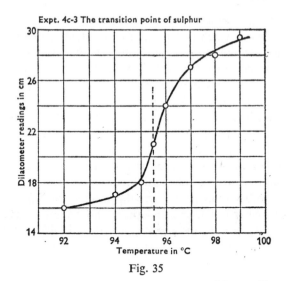

Fig. 35

beaker. Fit a rubber stopper carrying a wide capillary tube and scale, as shown in the diagram, taking care to see that there are no air bubbles in the test-tube. (This process requires considerable patience.) Slowly raise the temperature, keeping the heating bath well stirred. The rate of heating should be very slow, so as to give the sulphur time to rearrange itself in its new crystal pattern at the transition temperature. Note the position of the acid in the capillary every degree of temperature from 90 to 100°C. Plot the scale readings against the temperature. Fig. 35 shows some results obtained in an experiment using this method and apparatus, the transition occurring at about the correct temperature, 95·5°C.

LIQUID CRYSTALS

Expt. 4c-4

Prepare a little cholesteryl benzoate as follows: Boil about 1 g of cholesterol with 2–3 cm³ of benzoyl chloride for some 2 or 3 min, and pour the liquid into about 20 cm³ of cold alcohol. On standing, crystals of cholesteryl benzoate separate out. Filter off and dry them. Place some of the dry crystals in a melting point tube attached to a thermometer, and heat in a bath of medicinal paraffin. The cholesteryl benzoate should melt to a turbid liquid at 146°C and to a clear liquid at 178°C.

The term 'liquid crystal' was first used to describe the behaviour of ammonium oleate, which is deposited from alcoholic solution in what appear to be definite 'crystalline' forms. However, the 'crystals' are somewhat rounded and flow into one another when they touch. A number of substances are now known which exist in a state intermediate between liquid and crystalline solid, and which is called the 'meso-morphic state'. Examination of mesoforms is usually made by melting the substance between glass slides and observing the changes on a polarizing microscope between crossed polaroids.

In general, the mesoforms may be regarded as crystalline in two or one dimensions (the 'smectic' and 'nematic' states respectively) and liquid in the other dimensions, i.e. the atoms are linked into plates (smectic) or strings (nematic), and these slide freely over each other. The following is a list of substances which form liquid crystals, showing the transition temperatures: T_s the melting point of solid to smectic, T_n to nematic, and T_l to isotropic liquid:

	$T_s °C$	$T_n °C$	$T_l °C$
Thallous stearate	118	—	163
Ethyl p-azoxybenzoate	114	—	120
Cholesteryl benzoate	145	—	178
p-Azoxyanisole	—	93	150
Ethyl p-toluol-p'-amino-cinnamate	96	107	118

5 Phase relations

ONE-COMPONENT SYSTEMS

It will be convenient to review here the relationships between the various states—gaseous, liquid and solid—in which a pure substance can exist, and to introduce the ideas of Willard Gibbs, in which these relationships are summarized by means of phase diagrams. This will be a theoretical section, but without some explanation of the meaning of phase diagrams, the experimental work in some subsequent sections will be less easily understood. Once the phase diagram for a particular substance, or mixture of substances, has been drawn, it sums up the behaviour of the substance or mixture in a visual form that is readily interpreted. In this section, the construction of phase diagrams for two substances, namely, water and sulphur, will be discussed.

These considerations apply to the substance or substances in a *closed system*, i.e. in the absence of other substances. It is often convenient to think of the substances as being contained in a cylinder fitted with a piston on which the pressure can vary, the whole being immersed in a bath whose temperature can be controlled.

We have seen that when the temperature of a pure substance is raised, the substance may pass from the solid through the liquid to the gaseous state. The temperatures at which it changes from one state to another are not fixed but depend on the external pressure. It is very convenient to bring the data relating to these changes together on one diagram, and such a diagram is called a 'phase diagram'. The different physical forms, i.e. solid, liquid and vapour, are referred to as 'phases'. *A phase* is defined as 'a homogeneous part of a system bounded by a surface'.

The diagram is sometimes called an equilibrium diagram. The diagram for water is shown in Fig. 36 (not to scale).

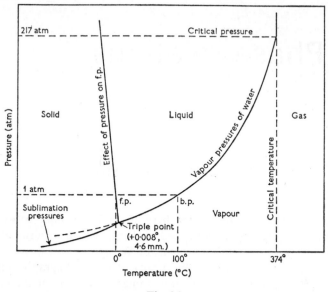

Fig. 36

For those substances which exist in more than one crystalline form (each of which constitutes a separate solid phase), the phase diagram is more complex. Sulphur is an example, and its phase diagram is shown in Fig. 37 (not to scale).

The areas S_α and S_β indicate the range of conditions for the existence of rhombic and monoclinic sulphur respectively. QN is the vapour pressure curve of liquid sulphur. OQ is the sublimation pressure curve of monoclinic sulphur. LO is the sublimation pressure curve of rhombic sulphur.

When rhombic sulphur, in equilibrium with its vapour, is slowly heated, the curve LO is followed. At 96°C the rhombic sulphur changes into monoclinic sulphur, and this temperature is the transition point, O (see p. 79). This is the only temperature at which both solid forms can coexist with their vapour. When the temperature is raised to 118°C the monoclinic sulphur melts (Q). Both the transition point and the melting point are affected by the external pressure, and the effect of pressure is represented by the curves OP and QP respectively. Since the transition point is more affected by increase of pressure than is the melting point, these curves meet at the point P, where $t = 151$°C. Above this temperature monoclinic sulphur cannot exist at all. There

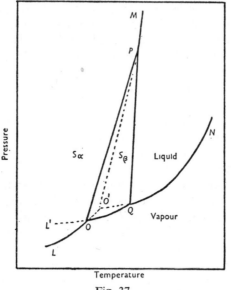

Fig. 37

is thus only a limited range of conditions of temperature and pressure under which this allotrope of sulphur is stable. The pressure at the point P is 1290 atmospheres, so the slopes of OP and QP have had to be greatly exaggerated in the figure.

SUPERCOOLING AND METASTABILITY

In addition to the conditions under which sulphur can exist as stable phases, it is also possible to obtain the substance under other conditions, when it is, however, not stable, but metastable. Thus, if monoclinic sulphur above 96°C is suddenly cooled, its vapour pressure is represented by points on the curve OL', an extension of QO. Its vapour pressure at any particular temperature is then greater than that of the rhombic, stable form, to which it slowly reverts on standing.

Similarly, liquid sulphur may be supercooled, i.e. cooled below its freezing point, without solidification. It is then in a metastable state; its vapour pressure will be higher than the solid form at the same temperature, and it will easily solidify.

There is thus a similarity between a transition point and a melting point; they are both temperatures at which two phases of the substance are in equilibrium, two solid phases in the former case, and a solid and a liquid phase in the latter; and they are both temperatures beyond which one phase may cool without undergoing change into the other. The metastable condition so produced is more easily upset in the case of a supercooled liquid than of a supercooled solid. This would be expected from the kinetic view of matter, for the metastable substance possesses potential energy (stored as vibrational energy of the constituent atoms), which it loses on transformation into the stable form. The moving atoms of a liquid will take up their positions in the crystal of the corresponding solid more easily than the vibrating, but anchored, atoms of one crystalline form will adopt new positions in the other crystal structure.

Rhombic sulphur can also be obtained in a metastable state. If it is *rapidly* heated, it does not become monoclinic sulphur at 96°C, but follows the extension of LO to O'. The temperature at O' is 113°C, and is the melting point of rhombic sulphur. QO' represents the vapour pressure of supercooled liquid sulphur, and PO' represents the effect of pressure on the melting point of rhombic sulphur.

It should be remembered that the diagram is simply a concise representation of experimental measurements, and forms a summary of the behaviour of the substance. Once such a diagram has been drawn, it is of great use in further work on the substance in giving information about the conditions under which the various phases are stable. Thus, it is clear from the sulphur diagram (i) that the four phases, vapour, liquid, S_α and S_β, can never exist together in equilibrium, (ii) that three phases can coexist but only under certain conditions (given by the points O, P and Q), and that if either temperature or pressure is arbitrarily changed, two of the three phases will disappear, but that if both are changed to a point on a line, one phase only will disappear; and so on.

ENANTIOTROPY AND MONOTROPY

A substance such as sulphur, whose two forms can both be obtained in a stable state under suitable conditions, is said to be *enantiotropic*. The polymorphs of an enantiotropic substance can be changed reversibly one into the other by appropriate changes of temperature (see p. 80). There are other substances, of which phosphorus is an example, of whose polymorphs this is not true. Thus white phosphorus is always metastable and changes slowly into red phosphorus. This process cannot be reversed by simple changes in temperature and pressure. White phosphorus is obtained by distillation, when the rapid cooling of vapour and liquid produces the metastable form. For these *monotropic* substances, the sublimation pressures of the metastable polymorph are higher than those of the stable form at all temperatures, see *L'O'* in Fig. 38.

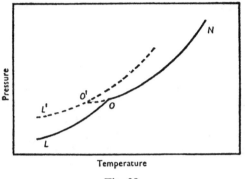

Fig. 38

THE PHASE RULE

From the phase diagrams of water and sulphur we can see that there is some relationship between the number of phases that can exist together in equilibrium and the range of conditions under which this is possible. Thus water can exist as one phase, say liquid, under a wide range of both temperature and pressure without vaporizing or solidifying. It can also exist as two phases together, say liquid and vapour, over a long temperature range, but at any particular temperature the vapour will exert a certain definite pressure. If the pressure is changed at this temperature, one of the phases will disappear, i.e. either all the vapour will condense to liquid, or all the liquid will evaporate into vapour. Finally, water can exist as three phases in equilibrium, but only at 0·0076°C and 4·57 mm pressure. It is thus clear that as the number of phases in the system is increased by one, the number of conditions that can be independently changed without altering the number of phases present is decreased by one.

Systems consisting of one substance only, e.g. water or sulphur, are known as 'one-component systems'. The substance may exist in several forms or phases, but each phase is made up of nothing but the one substance or *component*. Systems of two, three and more components will be considered later. In a one-component system, the conditions that can be altered to vary the state of the system are only two, i.e. temperature and pressure, but in systems of more components, other quantities can be varied, namely, the relative concentrations of the components in each phase. For example, in a system consisting of a salt and water, the possible variables are temperature, pressure and the concentration of the salt in the water. Again, the number of conditions that can be independently changed without altering the state of the system depends on the number of phases present. Thus, at any particular temperature, dilute solutions of a variety of strengths can be obtained, but a saturated solution of only one particular strength can have a stable existence.

The number of variables of condition that can be independently changed without altering the number of phases present, is called the number of *degrees of freedom* of the system. The phase rule expresses the number of degrees of freedom in terms of the number of components and the number of phases present in equilibrium. The relation was deduced from the laws of thermodynamics by Willard Gibbs in 1874, and states that $F = C - P + 2$, where F is the number of degrees of freedom, C is the number of components in the system and P is the number of phases present in equilibrium.

The method of using the rule is usually as follows:

(1) Experimental observations establish the number of forms in which the various components are capable of existing.

(2) The phase rule limits the number of phases that can exist together in equilibrium under specified conditions.

(3) Further experiments then provide the quantitative data for drawing up the phase diagram.

SOME EXAMPLES OF ONE-COMPONENT SYSTEMS

Water

Water can exist in three phases under ordinary conditions, i.e. one gaseous, one liquid and one solid. (At low temperatures and high pressures several other solid forms exist.) The phase rule gives the following information:

(*a*) When only one phase is present, there are $C - P + 2 = 2$ degrees of freedom; the system is said to be 'bivariant'.

(*b*) When there are two phases present, there is 1 degree of freedom; the system is 'univariant'.

(c) When there are three phases present, there are no degrees of freedom; the system is 'invariant'.

We have seen that water existing as one phase only is represented by an area on the phase diagram. The two variables of temperature and pressure can both be changed within limits without causing a change in the number of phases.

If water exists as two phases in equilibrium, e.g. liquid and vapour, only one variable can be arbitrarily changed; the other is a 'dependent variable'. Thus, by fixing, say, the temperature only, we can completely define the system. If an attempt is made to change the pressure of the vapour without changing the temperature, one of the phases will disappear, i.e. if the pressure is increased the vapour will liquefy. Likewise, if the temperature is raised, the pressure remaining constant, the liquid will vaporize. The conditions of temperature and pressure under which water will exist as two phases in equilibrium are represented on the phase diagram by points lying along three lines.

Finally, the three phases can only exist together at 4·57 mm pressure and 0·0076°C, conditions represented by the so-called 'triple-point' on the diagram.

The phase rule thus tells us that the phase diagram will consist of three areas separated by three lines meeting at a point, and it then remains to determine by experiment the quantitative data with which to construct the diagram.

Sulphur

Sulphur can exist in four stable phases, vapour, liquid and rhombic and monoclinic crystals. The phase rule tells us that for $P = 1$, $F = 2$; for $P = 2$, $F = 1$; for $P = 3$, $F = 0$. There are no conditions under which all four phases can exist together in equilibrium. The diagram will consist, therefore, of four areas, six lines and three invariant points.

TWO-COMPONENT SYSTEMS

(a) A gas and a liquid

Expt. 5a-1

Partly fill a beaker with soda water from a siphon and note the appearance of bubbles of gas. Warm the solution and observe the formation of more bubbles.

The solubility of gases in water varies enormously from one gas to another; for example, ammonia is 60,000 times as soluble as hydrogen by volume. Gases may be divided into two groups, those in one group being much more soluble than those in the other. In general, the more soluble gases react chemically with the water, whereas the less soluble gases do not. The latter include the 'permanent gases', oxygen, hydrogen, nitrogen, the inert gases, etc.; the more soluble group contains the more easily condensable gases, ammonia, hydrogen chloride, sulphur dioxide, etc.

The above experiment shows that the solubility of a gas is decreased by lowering the pressure and by raising the temperature. From the point of view of the phase rule, an equilibrium between a liquid and a gas (which does not react chemically with the liquid) has two degrees of freedom, for such a system consists of two components and two phases ($F = C - P + 2$). The three possible variables are: (i) temperature, (ii) pressure and (iii) the solubility of the gas. The existence of two degrees of freedom indicates that, at a fixed temperature, the solubility depends upon the pressure, or again, for a fixed pressure, the solubility varies with the temperature. This is as far as the phase rule takes us—the manner in which one variable varies with another has to be determined by experiment. Measurements show that, at a given temperature, the mass of gas dissolved by a certain mass of water (i.e. the solubility) is, for many gases, directly proportional to the pressure, $m = kp$. This relation is known as Henry's Law. The law is followed by the less soluble gases, but not by gases which react with water. The observation that a gas does not follow Henry's law may be useful in indicating that the substance does not exist in the same molecular form in solution as in the gaseous state. This is illustrated by the following experiment (see also p. 20).

Expt. 5a-2　*To determine the solubility of ammonia in water*

Prepare a small U-tube shaped as shown in Fig. 39. Clean and weigh it, then half fill it with 0·880 ammonia. Connect it to a small flask also containing 0·880 ammonia. Surround the U-tube with a beaker of water kept at the temperature at which the solubility is being measured. Warm the flask and allow ammonia gas to bubble through the U-tube for about 10 min. Then disconnect the flask and seal off the U-tube at *A* and *B*. Weigh the tube containing the saturated solution of ammonia, together with the two pieces of glass removed in the sealing.

Put 50 cm³ of sulphuric acid of known concentration (about M) in an evaporating dish, place the U-tube in the dish and add distilled water until the tube is completely submerged. Hold the bulb under the acid and break one seal. When most of the ammonia has been absorbed, break the bulb with a pestle. Add a little methyl orange

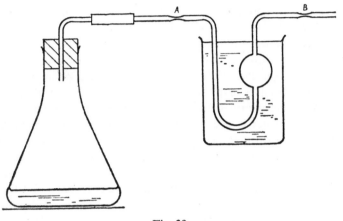

Fig. 39

and titrate the remaining acid with standard sodium carbonate solution. Calculate the weight of ammonia in the saturated solution, and by subtraction, obtain the weight of water. Hence calculate the solubility of ammonia in grams per 100 g of water at atmospheric pressure and at the temperature of the experiment. If time permits, repeat the determination at different temperatures. The solubility of ammonia in water at 12°C is about 63 g per 100 g of water.

(b) Two liquids

MUTUAL SOLUBILITY

Expt. 5b-1

Examine the miscibility of the following liquids with (*a*) water, (*b*) benzene, by shaking small quantities together in a test-tube: ethyl alcohol, glycerine, acetic acid; nitrobenzene, aniline, ether, ethyl acetate.

The above liquids may be divided broadly speaking into two classes —those that mix with water and those that do not. In general, the latter will mix with benzene, and members of each class will mix with each other. It can be seen from the above list that those liquids that mix with water all contain a hydroxyl group, and many of those of the other class are structurally similar to benzene. Thus, similarity of structure appears to be related to miscibility. (It is interesting to recall that this is true, not only of liquid, but also of solid solutions.)

Water and benzene differ in several other respects, and these other differences may throw light on their dissimilarity as solvents. (i) Water is an ionizing solvent. Although water itself is a poor conductor of electricity, it dissolves many substances to form solutions that are good conductors. Benzene does not behave in this way; thus a solution of acetic acid in water is a conductor, whereas a solution of acetic acid in benzene is not. (ii) Water has a high dielectric constant and benzene a low one. This means that water molecules have a high dipole moment, i.e. the molecule may be regarded as a small body in which there is a non-uniform distribution of electric charge, so that there is a difference of potential between the two 'ends' of the molecule. Such molecules will tend to attract each other strongly. (iii) Water has an exceptionally high boiling point for its formula weight and does not follow Trouton's rule (see Chap. 3c). This is due to the existence of the dipole moment which results in an aggregation of the molecules. When crystals of ice melt, they do not give rise to a fully disordered mixture of H_2O particles, but to an assemblage which may be regarded as retaining something of the structure of ice. The water molecules are held together by hydrogen bonds which are formed between the hydrogen atom of one water molecule and the oxygen atom of another (see Fig. 40). This attraction of the water molecules for each other helps to explain why water will not mix with a liquid such as benzene. When the two liquids are shaken together, the benzene molecules are 'elbowed out' by the water molecules, whose mutual attraction brings them together, and the benzene is forced to form a separate phase.

Some liquids, such as alcohol, mix with water in all proportions, but others are only partially soluble, e.g. water dissolves ether only up to about 7% at room temperature. With increase in temperature, ether becomes less soluble in water, but the solubility of water in ether increases. The solubility of phenol in water and of water in phenol both increase with rise in temperature. The solubility curves approach and meet at a temperature of 66·5°C and a composition of 33% phenol.

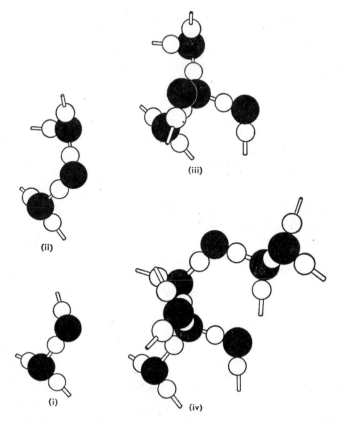

Fig. 40. Hydrogen bonding of water molecules into tetrahedral structures.

This temperature is known as the 'critical solution temperature'; above it, the two liquids mix in all proportions. Mixtures of aniline and water also exhibit a higher critical solution temperature (167°C). Some pairs of liquids, however, become more soluble in each other as the temperature falls, and exhibit a lower critical solution temperature. Dimethylamine and water, paraldehyde and water, behave in this way.

The critical solution temperature is extremely sensitive to small quantities of impurities. Thus, 1% of water raises the value for ethyl alcohol and petroleum by 17 deg. The change in the critical solution temperature is a measure of the amount of impurity present. By determining the critical solution temperature for petrol and aniline, the aromatic impurity in the petrol has been estimated. The dissolved substances in urine can be estimated from the critical solution temperature with phenol. 'A kidney producing urine that gives a critical solution temperature rise of 8° is in good order, and is functioning exceedingly well if the rise is 11–16°' (S. T. Bowden, *The Phase Rule and Phase Reactions*, p. 104).

Expt. 5b-2   The critical solution temperature for phenol and water

Make a slight constriction near the open end of each of eight clean 6 in. test-tubes. Weigh one tube, and weigh into it 1·0 g of dry phenol. Introduce 4 cm³ of water, weigh again, and seal off at the constriction. Prepare similar tubes containing the following weights:

Phenol (g)	0·5	1·0	2·0	2·5	3·0	3·25	3·5
Water (g)	4·5	4·0	3·0	2·5	2·0	1·75	1·5
Phenol percentage	10	20	40	50	60	65	70

Specimen result

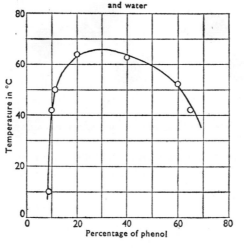

Fig. 41

Immerse one of the tubes in a large beaker of water and gradually raise the temperature. Shake the tube vigorously and note the temperature at which the turbid liquid becomes clear. Allow the water-bath to cool, and note the temperature at which the well-shaken liquid goes cloudy again. The required temperature may be obtained most accurately by adjusting the temperature of the bath so that the turbid liquid will just not clear and, if cleared by gentle warming over a flame, will not go cloudy on re-immersion in the bath.

Repeat the determination with the other tubes, and plot temperature against percentage composition. It may be found desirable to make another mixture or two to complete or extend the graph.

VAPOUR PRESSURE: (i) IMMISCIBLE LIQUIDS

Expt. 5b-3

Determine the boiling point of 'mixtures' of two immiscible liquids (e.g. chlorobenzene and water) in various proportions. Suspend a thermometer with its bulb immersed in the 'mixture' boiling in a flask. Note that the boiling point is lower than that of either liquid alone and is independent of the relative proportions in which the two liquids are present.

The vapour pressure of two immiscible liquids together is equal to the sum of the vapour pressures exerted by the two liquids separately at the same temperature; the vapour pressure of the two liquids together is thus independent of the relative amounts of each present. This follows from the phase rule, for there are three phases and two

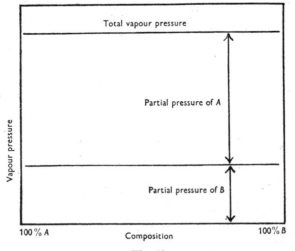

Fig. 42

components and therefore $2 - 3 + 2 = 1$ degree of freedom. Hence, at a given temperature, the system is invariant, i.e. there is only one value for the vapour pressure, no matter what the relative amounts of the components may be. The vapour pressure–composition diagram is shown in Fig. 42.

Steam distillation

If steam is passed through a 'mixture' of water and a liquid with which it is not miscible, such as chlorobenzene, the 'mixture' boils at a temperature where the sum of the vapour pressures of water and chlorobenzene becomes equal to the external (atmospheric) pressure. This temperature is, of course, lower than the boiling points of either liquid. In this way, the whole of the chlorobenzene can be distilled at a conveniently low temperature. The method is of great use in purifying liquids from non-volatile impurities, particularly those liquids that decompose at temperatures near their boiling points.

The molecular weight of the liquid that is being steam-distilled can be found if the relative weights of water and the other liquid in the distillate are measured. This ratio will be equal to the ratio of the weights of the two substances present in the mixed vapour before it is condensed to the distillate. By Avogadro's hypothesis, the ratio of the partial pressures of the two vapours, p_1/p_2, will be equal to the ratio of the numbers of molecules, n_1/n_2, and the latter can be expressed in terms of the weights w_1 and w_2, and the molecular weights m_1 and m_2. Thus

$$\frac{p_1}{p_2} = \frac{n_1}{n_2} = \frac{w_1}{m_1} \bigg/ \frac{w_2}{m_2}.$$

So if we know the vapour pressures of the two liquids at the boiling point of the mixture, and the molecular weight of one, we can find the molecular weight of the other.

Expt. 5b-4   Determination of the molecular weight of chlorobenzene by steam distillation

(1) Fit a 500 cm³ flask with a cork carrying a 100 deg C thermometer reading in 0·1 deg, and two delivery tubes, one extending to the bottom of the flask. The thermometer bulb should be immersed in the liquid. Place about 40 cm³ of chlorobenzene and 100 cm³ of distilled water in the flask and pass steam through the mixture, condensing the distillate by means of a condenser in the usual way.

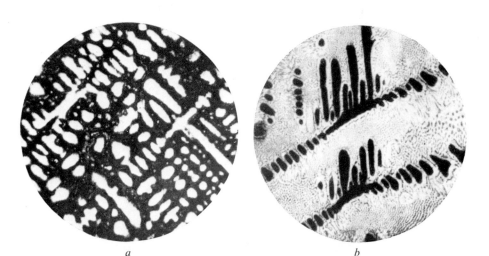

a   *b*

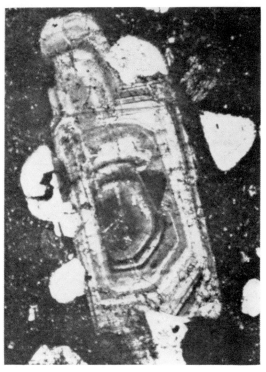

c

6 SOLID SOLUTIONS

a 50% copper-nickel alloy, etched to show dendritic structure (×86)
b Silver-copper alloy showing dendrites of copper (containing a little silver in solid
     solution) lying in a silver-copper eutectic (×200)
c Zoned plagioclase felspar (solid solutions of sodium and calcium alumino-
     silicates) (crossed nicols, ×50)

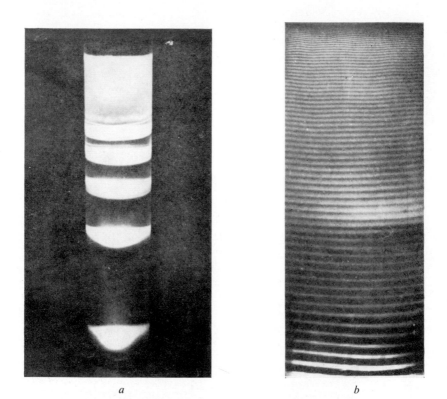

7 LEISEGANG RINGS

a Magnesium hydroxide bands in gelatine ($\times 4/5$)
b Silver chromate bands in gelatine ($\times 3/2$)
c Banded agate ($\times 1$)

Note the temperature of the mixed liquids when the distillation is occurring. Collect the distillate in a measuring cylinder and read off the volumes of chlorobenzene and water after about three-quarters of the chlorobenzene has distilled over. From the densities of the liquids, calculate their weights; alternatively, separate the liquids by using a separating funnel and weigh them. If the separation is difficult owing to the relatively small volume of water, add a known quantity of water, say 20 cm^3, and subtract from the measured volume. Look up the vapour pressure of water at the temperature registered by the thermometer, and subtract this pressure from the atmospheric pressure to obtain the vapour pressure of the chlorobenzene. Calculate the molecular weight of the chlorobenzene as indicated in the preceding paragraph. A result within 4% of the true value (112·5) can be obtained.

(2) Use this method of steam distillation to extract the oil from some fruit or seed and to find its molecular weight. A suitable example is orange peel, which yields an oil, the principal component of which is limonene, $C_{10}H_{16}$.

Finely chop or mince the orange part of the skins of about 20 oranges. Cover them with water in a large distilling flask, and steam distil until little more oil distils. Use a separating funnel to separate the oil. Measure the relative weights of oil and water and calculate the approximate molecular weight of the oil. Dry it over anhydrous calcium chloride and measure the boiling point. (Limonene boils at 176°C.) Attempt to find out how many unsaturated double bonds there are in the molecule by titration with a solution of bromine. Make a solution of known molarity of the oil in benzene, measure out 10 cm^3, and run in a solution of bromine in benzene, also of known molarity. How will you determine the end-point?

(ii) MISCIBLE LIQUIDS

Expt. 5b-5

Put some ethyl alcohol into a burette and some distilled water into another. Run 5 cm^3 of the alcohol into each of four test-tubes and add respectively 1, 2, 4, 6, cm^3 of water. Determine the boiling points of the mixtures by suspending the test-tubes in a large beaker containing a little boiling water, and noting the steady temperatures recorded by thermometers with their bulbs immersed in the boiling liquid mixtures.

It is clear from the results of this experiment that the boiling points of mixtures of two liquids which mix together in all proportions, unlike those of two immiscible liquids, vary with the composition of the mixture.

The vapour pressure of a given mixture of liquids will be equal to the sum of the partial vapour pressures of the constituents. Raoult assumed that, in the vapour above a mixture of liquids, the partial vapour pressure of each constituent is proportional to its molar fraction in the liquid. Thus if p_a is the vapour pressure of pure liquid A at a certain temperature and if p_a' is its partial vapour pressure above a mixture containing n_a mols of A and n_b mols of B, then

$$p_a' = p_a \frac{n_a}{n_a + n_b} \quad \text{and} \quad p_b' = p_b \frac{n_b}{n_a + n_b}.$$

The vapour pressure of the mixture $p_a' + p_b'$, will vary linearly with the composition of the liquid expressed as a molar fraction, for it follows that

$$p_a' + p_b' = (p_a - p_b) \frac{n_a}{n_a + n_b} + p_b.$$

When one constituent, say B, is non-volatile, i.e. $p_b = 0$, Raoult's law takes the form

$$\frac{p_a - p_a'}{P_a} = \frac{n_b}{n_b + n_b}.$$

Liquids that obey Raoult's law and behave in this way are called 'ideal liquids'. Real liquids, however, diverge from the law, and the

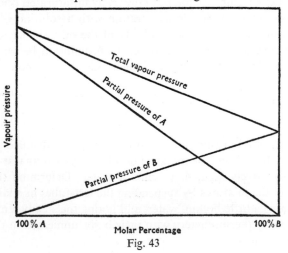

Fig. 43

vapour pressure-composition graphs for their mixtures are not straight lines, but curves of the three types illustrated in Fig 44. In the first type, the vapour pressures of all the mixtures have values lying between

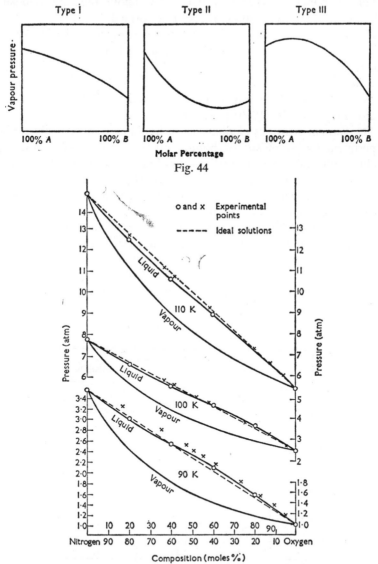

Fig. 44

Fig. 45. Pressure-composition diagrams for mixtures of liquid oxygen and liquid nitrogen

those of the pure liquids; in the second type, addition of either component to the other lowers the vapour pressure, hence a mixture of minimum vapour pressure exists; and in the third type, addition of either component to the other raises the vapour pressure, and so a mixture of maximum vapour pressure is formed.

Deviations from Raoult's law might be expected to be least when the liquids consist of simple molecules having but little effect on each other. Figure 45 shows results obtained for mixtures of liquid oxygen and liquid nitrogen, and it can be seen that the experimental points lie very near the line drawn for ideal liquids. (Reference: F. Din, *Trans. For. Soc.*, 56 (1960) p. 678.)

The vapour pressures of mixtures of liquids are most easily studied experimentally by means of boiling point measurements. Since the boiling point of a mixture of liquids is that temperature at which the total pressure becomes equal to the external pressure, there is a close connexion between the vapour pressures of the liquid at various temperatures and its boiling points at various pressures. The phase diagram showing the boiling points of mixtures of varying composition might be expected to have similar characteristics to the diagram showing the relation between vapour pressure and composition. The temperature-composition diagrams for mixtures of two liquids are found, then, to reflect the pressure-composition diagrams, and to belong to the three types mentioned already.

Type 1

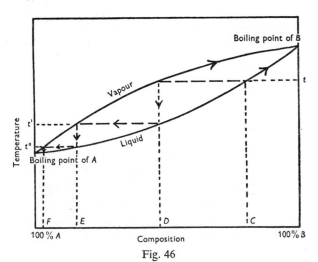

Fig. 46

Figure 46 shows the boiling points of mixtures of two miscible liquids of various compositions at some fixed external pressure. It may also be regarded as showing the temperatures at which liquid mixtures of various compositions can exist in equilibrium with their vapours at this pressure. If the composition of the vapour in equilibrium with the mixture C is determined, it is found to differ from that of the liquid, being richer than the liquid in the more volatile constituent. (This determination may be made by allowing the liquid to boil and analysing the first distillate, which will be identical in composition with the vapour from which it has been condensed.) The upper curve shows the composition of the vapour. Thus the vapour in equilibrium with the liquid mixture C at temperature t has a composition D, and so this will be the composition of the first distillate obtained when the mixture C is distilled. However, as the distillation proceeds, the mixture will change in composition and become increasingly rich in the less volatile component. The composition of the distillate will change correspondingly, as shown by the right-hand arrows in the figure.

Fractional distillation

The fact that the compositions of the liquid and vapour phases of a system of two miscible liquids are not the same, enables some separation of the mixture to be made. One distillation, in which the distillate is collected in separate 'fractions', will yield a range of samples containing the two constituents in varying proportions. A more complete separation can, however, be made by means of fractionating columns.

The liquid of composition C (Fig. 46) boils at temperature t, and produces a vapour of composition D. This vapour ascends the fractionating column and at first condenses on the cold plates. The latent heat thus released warms the plates, and liquid boils from one to the next until eventually it distils over at the top of the column. The composition of the liquid that condenses on plate 1 is D. When this liquid boils at temperature t', a vapour of composition E is produced, and so the liquid that condenses on plate 2 also has this composition. In this way the composition of the ascending vapour is progressively changed, until the vapour that finally leaves the column is almost pure A. The residual liquid thus becomes progressively richer in B, the less volatile constituent, and when all liquid A has been removed, liquid B eventually distils over. A complete separation such as this is not easy to obtain;

fractionating columns vary in their efficiency and various types have been designed to give as large a separation as possible, but with an ordinary plate, pear, or bead column, several fractionations are necessary to obtain a good separation.

Type 2

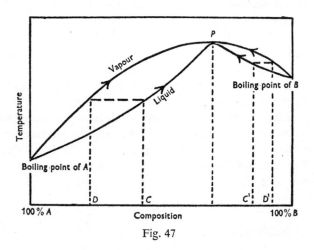

Fig. 47

Fractional distillation of a mixture of liquids that form a 'maximum boiling point mixture' separates them, not into the two pure components, but into one component and the maximum boiling point mixture. This 'constant boiling mixture' is not, however, a chemical compound, for (*a*) its composition varies with the pressure under which the mixture is distilled, and (*b*) the constituents are usually not in any simple molecular proportions.

Over a thousand mixtures of this type are known, examples of which are given below:

	Constant boiling mixture	
	Composition	b.p. (°C)
Water and hydrogen chloride (b.p. -80)	20·2% HCl	108·6
Water and hydrogen bromide (b.p. -73)	47·6% HBr	124·3
Water and hydrogen iodide (b.p. -35)	58% HI	127
Water and sulphuric acid	98·7% H_2SO_4	338
Water and formic acid (b.p. 99·9)	77% HCOOH	107

Since constant boiling mixtures cannot be separated into their constituents by distillation, the acids that form maximum boiling mixtures cannot be concentrated beyond the composition of the latter. Dalton knew, in 1802, that strong hydrochloric acid loses hydrogen chloride and becomes weaker on boiling, whereas dilute hydrochloric acid loses water and becomes stronger. The constant boiling mixture of hydrogen chloride and water, distilled under known external pressure, can be used in the preparation of standard solutions of the acid.

Type 3

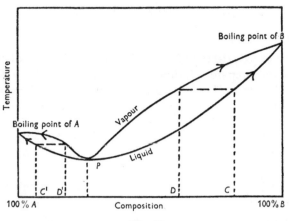

Fig. 48

Mixtures of this type are similar to type 2 in that fractional distillation separates them into one pure component and the constant boiling mixture, which in this case is a *minimum* boiling point mixture.

Water and ethyl alcohol (b.p. 78·3)   96% alcohol   b.p. 78·15°C
Water and n-propyl alcohol (b.p. 97·2)   71·7% alcohol b.p. 87·7°C
Water and pyridine (b.p. 115)   59% pyridine   b.p. 92°C
Benzene (80·2°) and ethyl alcohol (b.p. 78·3)   32·4% alcohol b.p. 68·2°C

Expt. 5b-6   The fractionation of a mixture of benzene and xylene

Distil a mixture of 50 cm³ of benzene (b.p. 81°) and 50 cm³ of laboratory xylene (b.p. 140°) in the ordinary way without using a fractionating column. Since mixtures of various compositions are characterized by their boiling points, the latter may be used as indications of the compositions of the distillates obtained. Therefore, collect the fractions that distil over temperature ranges of 10 deg, i.e. change the receiver when the temperature reaches 80°, 90°, 100°, 110° and so on. Measure the volumes of the fractions obtained.

Mix all the fractions together again, and repeat the experiment, this time using a fractionating column. Adjust the rate of boiling so that distillation occurs slowly. Again measure the volumes of the fractions that distil over 10 deg temperature ranges, and note that a better separation has been obtained. If the first, or last, fraction is redistilled, a further separation will be effected.

(c) A liquid and a solid

Expt. 5c-1

Dissolve exactly 10·0 g of potassium nitrate in 50 g of water and, by dilution, prepare solutions containing 3, 6, and 8 g per 50 g of water. Place test-tubes half-filled with the four solutions in a beaker of ice and salt, stir them well and note the temperatures at which solid first appears in them. The solid may be identified as ice or salt by noticing whether it floats or not. Enter the results on a graph showing temperature plotted against concentration of solution.

The formation of supersaturated solutions and crystallization from solution have been dealt with briefly in Chapter 4b. The relationships between solutes and solvents are conveniently described by means of the terminology of the phase rule, outlined at the beginning of this chapter. Systems of one component have been studied; mixtures of a salt with water, the subject of this section, are two-component systems. The number of variables is increased to three, namely, temperature, pressure and concentration. So in order to represent fully all the states of the system, a three-dimensional diagram would be required. It is usual to represent the effect of changing two variables at a time, the other being kept constant. Thus concentration-pressure diagrams are plotted at constant temperature, or, reference to the gas phase may be omitted and concentration-temperature diagrams plotted for the liquid and solid phases. In the latter case, the number of possible variables has been reduced by one, the pressure. The phase rule then becomes $F = C - P + 1$, and the system is referred to as a *condensed system*.

The data of Expt. 5c–1 form part of the phase diagram for the condensed system consisting of a salt and water. Expts. 5c–2 and 3 provide more complete data for the construction of the phase diagram shown in Fig. 49.

The line *CB* represents the solubility of the salt in water. Any point

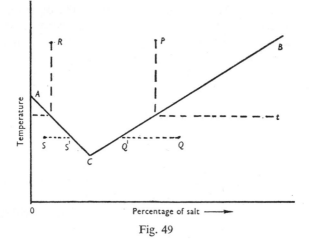

Fig. 49

on this line gives the solubility of the salt at some particular temperature, and represents the conditions of temperature and composition
under which solid salt and saturated solution can exist together in
equilibrium. The method of conducting Expt. 5c–2 stresses this latter
outlook. A point above *CB*, such as *P*, represents a dilute solution.
If this is cooled to the temperature *t*, it becomes saturated and solid salt
separates out. On further cooling the solution, it becomes poorer in salt,
and the temperature and composition of the solution will be represented
by points along *BC* in the direction of *C*. A point such as *Q* represents
a mixture of saturated solution of composition *Q'* with solid salt.

The curve *AC* is obtained from the results of Expt. 5c–3 and shows
the temperature at which solid first appears from solutions of various
concentrations as they are cooled. The solid that separates is shown
by its melting point to be pure ice. Hence the curve *AC* represents
conditions of temperature and composition under which ice and dilute
salt solutions can exist together in equilibrium. When a dilute solution
(say *R*) is cooled, ice begins to separate at some temperature *t*, and
this is the freezing point of the solution. If cooling is continued, the
solution becomes richer in salt, its freezing point falls, and the point
representing the temperature and composition of the solution moves
along *AC* towards *C*. A point such as *S* represents a mixture of solid
ice and a solution of composition *S'*.

The lines *AC* and *BC* meet at *C*, the temperature at which the residual
liquid from any solution that is being cooled will eventually solidify.
It is called 'the eutectic temperature' and the solid which separates is

called the 'eutectic'. Although it appears from the diagram that this solid has a definite composition, the eutectic is not a compound between the ice and salt, for it is not homogeneous, and can be seen under the microscope to consist of a mass of intermingled crystals of ice and salt. Its composition is not constant, but varies with pressure.

Expt. 5c-2 Measurement of the solubility of potassium nitrate in water at various temperatures

Weigh 15 g of potassium nitrate into a boiling-tube and add 10 cm³ of water from a burette. Raise the temperature rapidly to dissolve the salt. This strong solution will be used to make the others and no more weighing will be necessary. Allow the solution to cool and stir it with a thermometer. Note the temperature at which crystals appear. Add a further 2·5 cm³ of water and repeat the determination on this weaker solution. Continue in this way until 15 cm³ of water have been added, then add 5 cm³ at a time and increase to 10 and 20 at the lower concentrations. Plot the composition in grams dissolved in 100 g of water against the temperature of saturation.

Some inaccuracy is introduced through evaporation of the solvent in making the solutions, but this is small, as can be checked by analysing a weighed quantity of one of the solutions by evaporating it to dryness on a water bath in an evaporating dish and weighing the residue.

Expt. 5c-3   The freezing points of solutions of potassium nitrate

Use the apparatus described in Expt. 4a-1, but fill the conical flask with a freezing mixture of ice and salt kept at about −10°C. Find the freezing point of a solution of 5 g of potassium nitrate in 100 g of water. Stir the solution as cooling takes place and be prepared to 'seed' it with a crystal of ice if necessary. Note the temperature at which crystals first appear, and take this as the freezing point. Do not continue to cool the mixture, but allow it to warm up slowly, and note the temperature at which the crystals melt. Repeat the experiment, and this time continue the cooling until the whole mixture is solid. Notice the changes in temperature that occur. Allow the solid to warm up, and note the temperature at which the first sign of melting can be detected. This is the melting point of the solution, and differs from the freezing point. Repeat the whole process with solutions containing 8, 10 and 12 g of salt per 100 g of water. Enter the results on a diagram and combine it with that constructed from the results of the previous experiment.

Specimen results

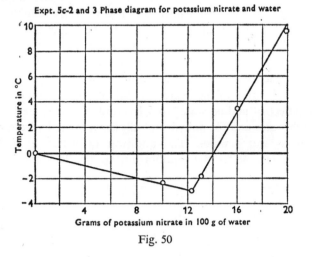

Expt. 5c-2 and 3 Phase diagram for potassium nitrate and water

Temperature in °C

Grams of potassium nitrate in 100 g of water

Fig. 50

Expt. 5c-4   Formation of compounds between a salt and water

Warm 150 g of pure *anhydrous* calcium chloride with 100 cm³ of water in a small beaker until the solid has dissolved. Allow to cool, stirring well, and note the temperature when solid first appears. Add 10 cm³ of water and repeat. Continue to dilute the solution and measure its freezing point in order to demonstrate that the latter passes through a maximum value in the neighbourhood of 30°C. If the appearance of solid is difficult to see, plot cooling curves and note the steady temperatures that are maintained when solid crystallizes out. The solid may be identified as ice or the salt hydrate by noticing whether it floats or not.

Specimen result

When a compound is formed between a salt and water, the phase diagram exhibits a maximum at the point corresponding to the composition of the hydrate. Part of the diagram for calcium chloride is shown in Fig. 51, and the above experiment is intended to provide data near both sides of the maximum.

AB represents the freezing point of dilute solutions of the salt. If a solution of composition P is cooled, the solid phase that separates is found to be $CaCl_2.6H_2O$. The liquid becomes poorer in salt, and eventually solidifies as the eutectic. If a solution, having the over-all composition corresponding to $CaCl_2.6H_2O$ is cooled, it solidifies completely at one temperature, C. Hence this is the freezing point and the melting point of the hexahydrate, and $B'C$ is its solubility curve. Addition of anhydrous calcium chloride to the hexahydrate lowers its freezing point, and CD represents this effect.

The presence of a maximum in the phase diagram shows that a compound is formed between the two components. The sharpness of the peak is an indication of the degree of stability of the compound. Consider the effect of adding water to a hydrate. The hydrate is in equilibrium with water and the anhydrous salt (or a lower hydrate): hydrate $\rightleftharpoons$ anhydrous salt + water. The addition of water to a relatively unstable hydrate shifts the equilibrium to the left, produces more

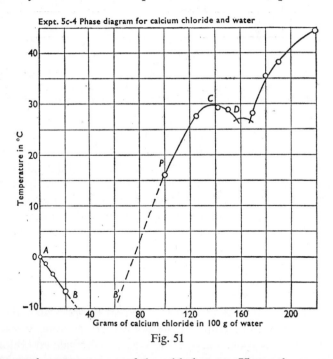

Expt. 5c-4 Phase diagram for calcium chloride and water

Fig. 51

hydrate, and removes some of the added water. Hence the composition of the liquid is not changed so much as in the case of a stable hydrate, where the equilibrium is already well over to the left, and most of the added water remains in the liquid phase. The equilibrium temperature is therefore lowered by a larger amount, producing a steeper curve for the stabler hydrate, as shown in Fig. 52.

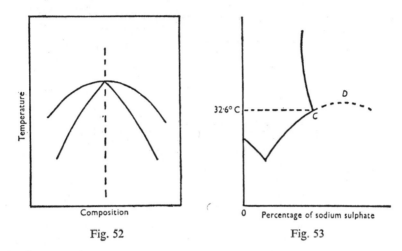

Fig. 52   Fig. 53

Sodium sulphate

The sodium sulphate-water diagram (Fig. 53) exhibits a sharp break, for the solubility curve of the anhydrous salt (note that the latter is more soluble at lower than at higher temperatures) meets the solubility curve of the decahydrate before the over-all composition of the system reaches $Na_2SO_4.10H_2O$. This simply means that the decahydrate decomposes at a lower temperature (C) than its melting point (D). The point C is the transition point of the anhydrous salt and the decahydrate, and can readily be determined by a thermometric method (see Expt. 5c–5).

EQUILIBRIUM DIAGRAMS FOR SALT HYDRATES AND THEIR VAPOURS

We shall see in Chapter 6 that the addition of a salt to water lowers the vapour pressure, and that the lowering is proportional to the molecular composition of the solution provided it is dilute. We shall now discuss the changes in the vapour pressure as the relative proportion of water to salt is reduced at constant temperature. The type of diagram obtained is shown in Fig. 54, in which the pressures of water vapour in equilibrium with copper sulphate and water in varying proportions are shown.

If water is removed from a dilute solution of copper sulphate, the

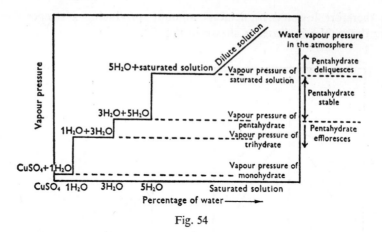

Fig. 54

vapour pressure falls continuously until the solution is saturated. If the proportion of water is now further reduced, the vapour pressure remains unaltered as long as some saturated solution is still present. When the composition of the system reaches that corresponding to the pentahydrate, the vapour pressure falls abruptly to a lower value. If more water is removed, thus causing some of the pentahydrate to be converted to trihydrate, the vapour pressure remains at this value as long as any pentahydrate is left. When the over-all composition of the system corresponds to the trihydrate, the vapour pressure again falls. In this way the 'stepped' graph shown is built up.

Let us now examine this from the point of view of the phase rule. In a dilute solution, $P = 2$ (1 liquid and 1 vapour phase). Hence $F = C - P + 2 = 2 - 2 + 2 = 2$. If we fix the temperature, there remains one degree of freedom; hence for every value of the concentration of the solution there is a fixed value for the remaining variable, i.e. the vapour pressure. But when the solution is saturated and is in equilibrium with the pentahydrate, P becomes 3 and F is reduced to 1. So when the temperature has been fixed, the system is invariant, and has a vapour pressure of one value only, no matter what are the relative amounts of solution and pentahydrate. When more water is removed so that the system consists of the penta- and trihydrates in equilibrium with water vapour, the system is again invariant at constant temperature. So the stepped shape of the diagram is to be expected from phase-rule considerations.

Efflorescence and deliquescence

Examples of salt hydrates that are stable in the air, of those that lose water or 'effloresce', and of those that absorb water or 'deliquesce', are well known. It is interesting to reflect that no hydrate would be stable in the atmosphere, but would either effloresce or deliquesce, if the composition-vapour pressure curve were smooth instead of stepped. Deliquescence can only occur when the partial pressure of water vapour in the atmosphere exceeds the vapour pressure of *the saturated solution of the salt hydrate*. If the water vapour pressure in the air is less than the vapour pressure of the dry hydrate, the hydrate will effloresce, losing water to the air and becoming the anhydrous salt or a lower hydrate of vapour pressure less than that of the atmospheric moisture. There is thus a range of value of the humidity of the atmosphere for which the hydrate will be stable and neither effloresce nor deliquesce (see Fig. 54).

The presence of steps in the vapour pressure-composition curve for mixtures of a solid and water indicates the formation of a compound between the two substances. This is sometimes the only way of establishing the existence or non-existence of such a compound. Thus the diagram for mixtures of silica and water affords no evidence for the existence of the many hypothetical silicic acids once invented to be the parents of the naturally occurring silicates. The diagram for sulphuric acid and water, however, confirms the existence of hydrates.

Expt. 5c-5 Determination of the transition point of sodium sulphate thermometrically

Sodium sulphate decahydrate loses its water of hydration above about 33°C; on cooling, the mixture evolves heat at the transition point and a steady temperature is maintained. Grind up some of the crystals and warm about a third of a test-tube of them to about 40°C, placing a thermometer reading to 0·1 deg C in the test-tube. Place the tube in a conical flask to protect it from draughts and take the temperature every minute as it cools. Plot a cooling curve and determine the transition temperature. A value within 0·1 deg C of 32·3°C should be obtained.

(d) Mixtures of two solids

Expt. 5d-1

Powder small quantities of m-dinitrobenzene (or naphthalene) (*A*) and *p*-nitrotoluene (*B*). Place a brass strip about 5 cm wide and at least 15 cm long on a tripod so that one end can be gently heated with a small flame. Place a little pile of substance *A* near the one end and a second pile alongside it consisting of *A* plus about 20% of *B*. Heat the other end of the strip until the piles melt and stir the second one well. Allow to cool; then warm up again slowly and note which pile melts first. Repeat the experiment with piles of pure *B* and *B* plus about 20% of *A*.

The melting point of a solid is, in general, lowered by the presence of small quantities of another solid. Moreover, the added impurity also causes the melting process to occur over a range of temperatures instead of sharply at one temperature as in the case of a pure solid. Thus the temperature at which the solid first begins to melt (the *melting point*) is different from that at which the whole mass is liquid. The latter temperature is called the *freezing point*; it is also the temperature at which solid first appears in the liquid when it cools. Hence the melting point is only identical with the freezing point for pure substances. (There are exceptions to this; see Figs. 55 and 58.) Familiar examples of solids which do not melt sharply (and are therefore mixtures) are candle-wax, butter, plumbers' solder, wrought iron, etc., all of which pass through a pasty stage before melting or solidifying completely.

Measurements of the melting and freezing points of two solids mixed in various proportions enable temperature-composition diagrams to be constructed. Such diagrams fall into three main groups, and in Expts. 5d, pairs of organic substances are chosen representing all three types. The first group contains substances that neither combine with each other nor form solid solutions (i.e. neither is soluble in the other in the solid state). In the second group are pairs of substances which are isomorphous and form solid solutions, while those in the third group combine with each other to form one or more compounds.

(1) EUTECTIC FORMATION

If two solids do not form compounds or solid solutions, the type of diagram shown in Fig. 55 can be drawn. Data obtained from mixtures of gold and thallium are given in the figure as an example. A is the melting point of pure thallium, B that of pure gold, AC is the freezing point curve of thallium containing increasing amounts of gold, and BC is the corresponding curve for alloys of gold containing increasing amounts of thallium. C represents the composition of the alloy which has the lowest freezing point of the whole series. This point, which defines both composition and temperature, is known as the *eutectic point*.

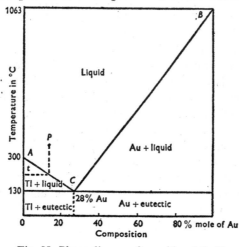

Fig. 55. Phase diagram for gold and thallium

As a melt of composition P is cooled, no solid appears until temperature t is reached, and the solid then formed is found to be pure thallium. As thallium continues to separate, the liquid becomes richer in gold. The temperature and the composition of the residual liquid are then represented by the line AC. At C the remaining liquid solidifies, forming the eutectic alloy. A section of the cast solid will show crystals of thallium embedded in a groundmass of the eutectic. The eutectic itself is not homogeneous, but consists of closely mingled crystals of thallium and gold. That the eutectic must be a mixture of two solid phases follows from the phase rule for point C is invariant, i.e. $F = 0$ and the system contains two components, hence the number of phases must be $C - F + 1 = 3$. One phase is liquid, therefore there must be two solid phases.

The following pairs of substances form this type of system:
Organic substances: naphthalene and p-toluidine; m-dinitrobenzene and azobenzene.

E B C—I

Metals: gold and thallium; lead and silver; tin and zinc.
Minerals: anorthite $(CaAl_2Si_2O_8)$ and diopside $(CaMgSi_2O_6)$.

(2) SOLID SOLUTIONS

In the previous section it was stated that when a melt of two substances that do not form solid solutions is cooled, the solid that crystallizes first is one of the pure components. However, if one substance is soluble in the other in the solid state, the first solid to separate from a melt of composition P as it cools, will contain both A and B and has the composition Q. The melt solidifies completely at temperature t' (Fig. 56). If a solid of composition P is heated, it begins to melt at this temperature and is not completely molten until the temperature t is reached. The lower curve is thus the melting point–composition curve and is called the 'solidus'. The upper curve is the freezing point–composition curve and is called the 'liquidus'.

The solid that crystallizes when a liquid melt of this type is cooled is homogeneous under the microscope and constitutes only one phase. This also follows from the phase rule for it is clear from Fig. 56 that, when the solid is crystallizing the system is univariant (being represented by a point on a line); hence $F = 1$, C is 2, therefore $P = C - F + 1 = 2$. There is one liquid phase, so there can only be

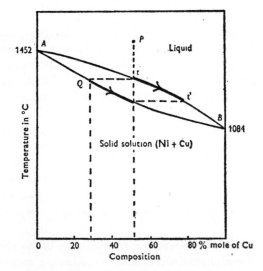

Fig. 56. Phase diagram for nickel and copper

one solid phase. This solid phase is called a *solid solution*, a term introduced by van't Hoff in 1890.

A solid solution differs fundamentally from a solution in a liquid, for a liquid has no rigid structure of its own into which the solute has to intrude, but the formation of a solid solution involves either the production of a new crystal structure including atoms or ions of both constituents, or, in certain cases, the accommodation of the solute particles in the interstices of the existing structure. For further details about the type of substances that form solid solutions see p. 78.

Refer again to Fig. 56. As the temperature falls below *t*, the liquid phase becomes richer in *B* and its composition is represented by points on the upper line or liquidus. The composition of the solid that crystallizes also alters. If the cooling is sufficiently slow, and the solid is always in equilibrium with the liquid, the composition of the solid follows the line *QB* until it reaches that of the original liquid *P* when solidification is complete. This means that the solid solution that first crystallized has been progressively changed in composition by diffusion of *B* into the solid phase, so that its composition varies continuously as the system continues to cool. It is only by means of this continual readjustment of composition that homogeneous solid solutions can be formed. Good examples are met in the study of naturally occurring minerals, e.g. the plagioclase felspars, which form a continuous series of solid solution of albite (sodium aluminium silicate, $NaAlSi_3O_8$) and anorthite (calcium aluminium silicate, $CaAl_2Si_2O_8$), and in metallurgy, e.g. copper and zinc in the brasses. However, if the melt does not cool slowly enough for the solid phase to maintain its equilibrium with the liquid, 'zoned crystals' form, which have a composition that varies from the core to the shell. Examples of these zoned crystals are found in rocks containing plagioclase felspars (Plate 6*c*), and in cast alloys, e.g. 50:50 copper-nickel. In the latter example, the zoned nature of the dendrites may be established by varying the time of etching of a micro-section; the outer layers are richer in copper and are therefore more readily etched than the cores, which appear light under the microscope (Plate 6*a*).

Solid solutions may also be formed in certain cases when one solid is placed in contact with the other and heated to a suitable temperature below the melting point. Diffusion of one substance into the other, atom by atom, occurs and the mixed crystal structure is built up. This is the basis of several industrial processes, e.g. sherardizing (the dissolution of zinc in the surface of iron) and case hardening (the dissolution of carbon in an iron surface).

Examples of systems that have phase diagrams of this type are: silver and gold, copper and nickel, tin and bismuth, and the alums.

It is interesting to note that, in these cases, one component *raises* the freezing point of the other.

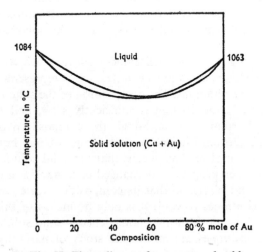

Fig. 57. Phase diagram for copper and gold

In the type of system shown in Fig. 57, the addition of either component to the other lowers the freezing point, and a minimum freezing point mixture is formed at some intermediate composition. The system composed of sodium and potassium carbonates is an example of this type. The mixture of equal parts of the two carbonates, which melt respectively at 820° and 860°C, is known as 'fusion mixture' and melts at 690°C. Other examples are: copper and gold, mercuric iodide and mercuric bromide.

Systems which form a maximum freezing point solid solution are extremely rare.

If there is a limit to the solubility of component B in component A and of A in B, i.e. the two components are not miscible in all proportions in the solid state, they are said to form a 'broken series of solid solutions', and a phase diagram such as that shown in Fig. 58 is obtained. When a melt of composition P is cooled, the composition of the solid that separates moves along QD and that of the residual liquid along *PC*. The more soluble the solid *B* is in solid *A*, the nearer *D* will be to *C*. The latter is a eutectic point, and the eutectic will consist of a mixture of two solid solutions, namely, those of compositions given by *D* and *E*. In the figure, the solid *A*, namely, silver, is only slightly

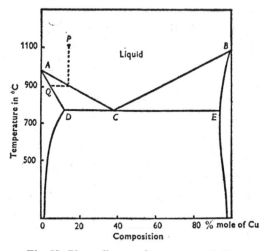

Fig. 58. Phase diagram for copper and silver

soluble in solid *B*, namely, copper, therefore the point *E* is some distance from *C*. Copper is rather more soluble in silver.

The diagram for the minerals orthoclase (potassium aluminium silicate) and albite (sodium aluminium silicate) is of this type, with a eutectic containing 60% of albite. The example quoted above is illustrated in Plate 6*b*, which shows dendrites of a solid solution of silver in copper lying in a groundmass of eutectic.

In plumber's solder, consisting of a mixture of tin and lead (33% tin), there is a wide gap between the solidus and the liquidus, indicating that the material remains in a pasty condition over a considerable temperature range. This is useful to the plumber as it gives him time to 'wipe' the soldered joint. Tinman's solder has a composition almost equal to that of the eutectic (63% tin), so that it sets sharply and does not go through a pasty stage.

(3) COMPOUND FORMATION

If the two components form a compound, the phase diagram shows a maximum at the composition of the compound. The states of the system represented by the various parts of the diagram are indicated in Fig. 59. It is most simply regarded as made up of two diagrams of type 1, i.e.

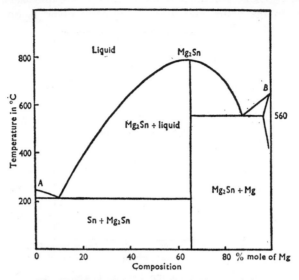

Fig. 59. Phase diagram for magnesium and tin

one for a system consisting of solid A and the compound of A and B, and the other of the compound and solid B. If several compounds are formed, the diagram shows a corresponding number of maxima, and the existence of maxima indicates the formation of compounds and enables their composition to be determined. This is often the only practicable method of detecting compound formation, particularly in the case of systems consisting of mixtures of metals. The sharpness of the maximum in the phase diagram is an indication of the stability of the compound. If the latter is so stable that it has no tendency to dissociate at its melting point, the maximum comes to a sharp peak.

Compounds between metals ('intermetallic compounds') are very numerous and present several points of interest:

(a) They are seldom formed between metals of the same group in the periodic classification.

(b) The ordinary valency rules are not followed, e.g. Cu_3Sn, $MgZn_2$, $PbMg$, $CuMg_2$, Al_3Mg_4, etc.

(c) They are usually hard and brittle, and relatively poor conductors of heat and electricity.

Examples: benzophenone and diphenylamine; p-chlorophenol and p-toluidine; copper and tin, Cu_3Sn; magnesium and tin, Mg_2Sn; magnesium and zinc, $MgZn_2$.

CONSTRUCTION OF TEMPERATURE-COMPOSITION DIAGRAMS

There are several methods available for obtaining the melting point–freezing point curves, of which two are described in the experiments below (see Expts. 5d-2 and 3).

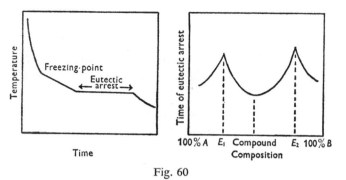

Fig. 60

In the first experiment, the temperatures at which the solid is first seen to melt and that at which melting is observed to be complete are recorded. This method is satisfactory for transparent liquids, but not practicable when the molten system is opaque.

In the second method, a cooling curve is plotted showing the variation of temperature as the liquid cools until it has completely solidified (Fig. 60). The rate of cooling alters sharply at the freezing point, and becomes slower as the solid crystallizes and releases its latent heat. The curve shows a second break when the eutectic point is reached, and the temperature then remains steady until all the eutectic has solidified and the whole mass is solid. The solid then continues to cool according to Newton's law. The horizontal portion of the curve which corresponds to the solidification of the eutectic is called the *eutectic arrest*. The length of the eutectic arrest depends upon the composition of the original melt, and is a maximum for a composition equal to that of the eutectic itself. The best way of determining the eutectic composition is to plot the length of the eutectic arrest obtained from a series of cooling curves against the composition of the melt. When the components of the system form a compound, the eutectic arrest curve shows two maxima corresponding to the compositions of the two eutectic mixtures, and a minimum at the composition of the compound (see Fig. 60).

Expt. 5d-2   The melting point method

This method consists in placing small quantities of the homogeneous mixtures in thin-walled capillary tubes, and observing the temperature at which melting begins and that at which it is complete. Naphthalene and β-naphthol are a suitable pair for this experiment. They give a continuous series of solid solutions. Weigh out accurately mixtures containing the following quantities:

| Naphthalene (g) | 3 | 3 | 3 | 2 | 1 |
| β-Naphthol (g) | 1 | 2 | 3 | 3 | 3 |

Melt the first mixture in a test-tube, shake, and cool it rapidly by immersing the tube in cold water. Remove the solidified mass,

Specimen result

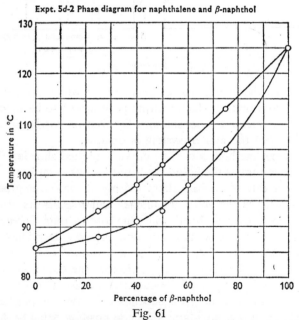

Expt. 5d-2 Phase diagram for naphthalene and β-naphthol

Fig. 61

powder it in a mortar, and transfer some to a melting point tube, made by drawing out a test-tube to a diameter of about 2 or 3 mm. Insert a fine glass rod as a stirrer. Attach the tube to a thermometer reading up to 150°C and heat it slowly and regularly in a bath of medicinal paraffin or a silicone oil. (Tri-cresyl phosphate is a useful liquid for a high temperature bath.) The temperature should rise not more than 1 deg per minute, and the bath should be well stirred. Note the temperature at which the solid first becomes sticky, which is when it is beginning to melt, and that at which the solid has just melted completely.

Repeat the determination for the other mixtures and for the pure components. Plot a graph of the melting and freezing points against the compositions of the mixtures (see Fig. 61).

Expt. 5d-3 The cooling curve method. Naphthalene and p-nitrotoluene

Place exactly 3 g of naphthalene in a test-tube and melt it by immersion in a beaker of boiling water. Insert a thermometer in the molten naphthalene and place the tube in a conical flask to protect it from draughts. Put some cotton wool in the top of the test-tube and in the neck of the flask. Read the temperature every minute and plot a cooling curve. The steady temperature gives the freezing point of the pure naphthalene.

Then add exactly 1 g of p-nitrotoluene, melt the mixture, mix by shaking gently, and plot a cooling curve as before. Note approximately the temperature when crystals first appear. Plot the curve as the readings are taken and continue until the flat part has been passed. The curve will show two changes of slope: the first corresponds to the first appearance of solid, and the second to the crystallization of the eutectic mixture.

Repeat the experiment using the following quantities:

Naphthalene (g)	3	3	3	1	1	2
p-Nitrotoluene (g)	1	2	3	9	3	3

The eutectic mixtures should crystallize at the same temperature from all the mixed liquids. Plot the freezing and melting points against the compositions of the mixtures (see Figs. 62, 63).

A suitable pair of substances to illustrate compound formation consists of p-toluidine (m.p. 43 °C) and α-naphthol (m.p. 94 °C). A compound is formed with a freezing point of 50 °C, and eutectics separate at 26° and 47°. A further example is provided by mixtures of urea (m.p. 132 °C) and phenol (m.p. 43 °C), which form a compound melting at 61 °C. Make up the mixtures in mole fractions. What do your results give for the molecular proportions of the two substances in the compound?

Specimen Results

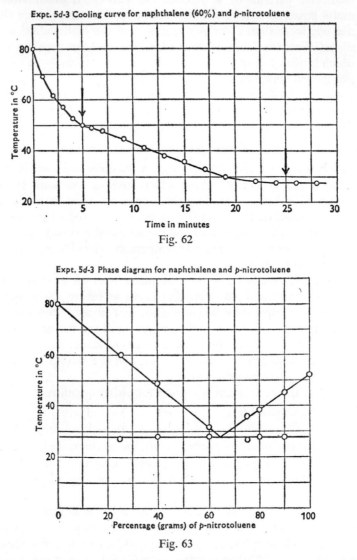

Expt. 5d-3 Cooling curve for naphthalene (60%) and p-nitrotoluene

Fig. 62

Expt. 5d-3 Phase diagram for naphthalene and p-nitrotoluene

Fig. 63

o-Chlorophenol and *p*-toluidine form another suitable pair, with eutectics at about 25° and 5°C. A compound is formed between equimolecular quantities and has a melting point of 39°C. (The presence of impurities lowers this temperature considerably.)

This method is not suitable for substances that form solid solutions; the method of Expt. 5d-2 above may be used for any type of mixture.

Expt. 5d-4   Phase diagram for tin-lead mixtures

The method consists in melting known weights of pure tin and lead in hard glass tubes and plotting cooling curves. Suitable compositions for the mixtures are: 100, 80, 70, 50, 30, 10% lead and 100% tin. Heat each mixture well above the melting point in a hard glass tube and place the tube in a fire-clay pot lined with asbestos-wool. Insert one junction of a thermocouple (made of fine iron and constantan or manganin wires) in the molten metal and connect the other ends of the wires to a millivoltmeter. The measurements on the pure tin and lead, which melt respectively at 232° and 326°C, provide a calibration for the thermocouple. Plot the temperatures against the time in order to obtain the cooling curves. These will show two changes of slope, the first when the excess component begins to separate, and the second when the eutectic solidifies. From the curves, construct an equilibrium diagram showing melting and freezing points plotted against composition. The diagram (Fig. 64) should show the formation of a eutectic containing about 63% of tin.

Extract the solid pellets of alloys from the test-tubes and prepare several for microscopic examination in the following manner. File a flat surface on the pellet, and then grind it on a series of emery papers of coarse, medium and fine grades placed on a flat glass surface. Suitable grades of emery paper are: 400, 500, 600, F, FF, 00, 0000. First rub the specimen backwards and forwards on the coarsest grade of paper, pressing firmly but not too hard. Then rub it on the next grade in a direction at right angles to the scratches produced by the coarser paper until these are no longer visible. Proceed in this way with the finer grades of emery. Finally polish the specimen with ordinary metal polish, using a soft cloth.

To reveal the structure, etch the surface by immersing the specimen in a mixture of 1 part concentrated nitric acid, 1 part glacial acetic acid and 8 parts of glycerine, warmed to about 40°C. The time needed will vary from 1 to 10 minutes and should be determined by trial. Remove the specimen from the etching solution. Wash it well, dry it in a hot air stream, or with paper tissues and examine it by reflected

Specimen result

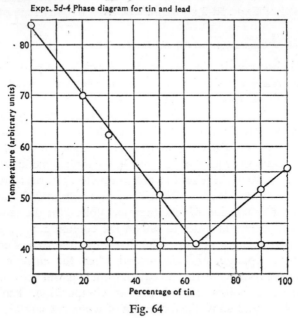

Expt. 5d-4 Phase diagram for tin and lead

Fig. 64

light under a low-power microscope. A ×20 objective with ×10 eye-piece is suitable. Place the specimen on a small block with its etched

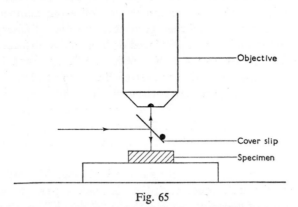

Fig. 65

surface horizontal. Light from a torch bulb is reflected on to it by a small glass coverslip fixed to a needle or pin by a spot of adhesive. The needle is mounted in a cork or wooden handle held in a clamp. Adjust the angle of the coverslip so that light falls on the specimen and is reflected back through the coverslip onto the objective lens of the microscope. Compare the appearances of the etched surfaces of the range of alloys. The 50 : 50 alloy for example, should show dark dentrites of lead lying in a lighter background of eutectic. Is the eutectic uniform or has it too got a structure?

(e) The distribution law

Expt. 5e-1

Make a few cm³ of a solution containing a very small crystal of iodine in benzene so that it has a good brown colour. Shake this in a small separating funnel with an equal volume of water containing a crystal of potassium iodide. Run the two layers into separate test-tubes and label them a_1 and b_1. Repeat the experiment with a benzene solution weaker in iodine, and label the layers a_2 and b_2. Make a quick and approximate comparison between the concentrations of iodine in a_1 and a_2 and in b_1 and b_2, by a simple colorimetric method thus: pour solutions a_1 and a_2 into flat-bottomed test-tubes of equal size until the depths of colour, seen on looking down the tube on to a white background, are equal. The concentrations of iodine in the two liquids will then be inversely proportional to the heights of liquid in the tubes. Repeat with b_1 and b_2 and note that the ratios a_1/a_2 and b_1/b_2 are the same. Hence $\dfrac{a_1}{b_1} = \dfrac{a_2}{b_2}$.

The mixture of iodine, benzene and water is an example of a three-component system. If we ignore the vapour phase, there are two phases, namely, a solution of iodine in benzene and a solution of iodine in water, and there are three variables, temperature and the concentrations of iodine in the two liquids. Therefore $F = 3 - 2 + 1 = 2$. Hence at any fixed temperature, only one of the concentrations can be arbitrarily chosen, the other is then fixed. The results of the above experiment bear this out, and show further that the concentration of the solute in one phase, c_1, is directly proportional to that in the other, c_2, viz. $c_1 = k \times c_2$. This relation is known as the Distribution Law, and the constant k is the Distribution Coefficient or Partition Coefficient for the solute distributed between the two given solvents, at a particular temperature.

Some systems appear not to follow the simple relationship, $c_1/c_2 = k$. Thus, measurements of the distribution of benzoic acid between water and benzene are represented by the relation $c_w = k\sqrt{c_b}$, where c_w and c_b are the concentrations of the benzoic acid in the two liquids. We shall see shortly that this result would be obtained if benzoic acid were associated as double molecules in the benzene but existed as single molecules in the water. It is found that similar apparent modifications of the relationship are necessary whenever the solute exists in different molecular forms in the two solvents, and Nernst formally stated the distribution law thus: 'If a dissolved substance is in the same molecular condition in both solvents, it possesses a distribution coefficient independent of concentration.'

It is interesting to compare the distribution law with the equilibrium law (see Chap. 12). They are similar in that they both relate the concentrations of substances that are in dynamic equilibrium with each other. They differ in that the equilibrium law applies to equilibria between different substances in the same phase, whereas the distribution law applies to equilibria across a phase boundary between one and the same molecular species present in both phases.

Association

Consider first a solute that exists as simple molecules in solvent 1, but partly as double molecules in solvent 2. Let c_2 be the total concentration of solute in solvent 2. If the degree of association is α, then the concentration of unassociated molecules in solvent 2 is $(1 - \alpha)c_2$, and that of associated molecules is αc_2. The equilibrium may be represented thus:

$$\begin{array}{c} c_1 \quad 2A \\ \hline \\ c_2 \quad \underset{(1-\alpha)c_2 \quad \alpha c_2}{2A \rightleftharpoons A_2} \end{array}$$

Applying the equilibrium law to the homogeneous equilibrium in solvent 2, we have $(1 - \alpha)^2 c_2{}^2 = K\alpha c_2$. Hence the concentration of unassociated molecules in solvent 2 is equal to $\sqrt{(K\alpha c_2)}$. The distribution law then states that $c_1 = k\sqrt{(\alpha c_2)}$. If the solute is almost completely associated in solvent 2, α is almost unity and the relation becomes $c_1 = k\sqrt{c_2}$ (see Expt. 5e-4).

Dissociation

If the solute is dissociated to a degree α in solvent 2, the concentration of undissociated molecules in that solvent is $c_2(1 - \alpha)$. The distribution law then states that $c_1 = kc_2(1 - \alpha)$. For small degrees of dissociation, this approximates to the normal form of the law:

$$
\frac{c_1 \quad AB}{\underset{c_2(1-\alpha)}{C_2 \quad AB \rightleftharpoons A + B}}
$$

Henry's law

Henry's law (see p. 90) may be regarded as a special case of the distribution law, where the solute, a gas, is distributed between one solvent and space. The concentration, c_1, in the former is the solubility, in grams per 100 g of solvent, and the concentration, c_2, in the latter, is the pressure of the gas, p. The distribution law states that $c_1/c_2 = k$, or $m = kp$, which is Henry's law.

EXTRACTION

The extraction of a solute from one solvent by means of a second solvent that does not mix with the first is often used in the preparation of organic substances. For example, aniline is extracted from aqueous solution by means of ether, in which it is more soluble and from which it is more easily obtained in the pure state. The ether is shaken with the aqueous solution and the aniline distributes itself between the two solvents. A fraction of the aniline, determined by the distribution coefficient, is thus transferred from the water to the ether. It is usually desired to extract as much aniline as possible without using too much ether and this is effected by using the ether in successive small quantities rather than all at once. That this procedure is more effective than one extraction may be demonstrated as follows.

Suppose M g of solute are dissolved in 100 cm³ of water and that 100 cm³ of ether are to be used. Suppose the distribution coefficient of the solute between water and ether is 3. Suppose first that all the ether is used at once. Let m_w and m_e be the masses of the solute in the water and ether respectively after extraction. Then, by the distribution law, $\dfrac{m_e}{100} = 3\dfrac{m_w}{100}$. Since $m_e + m_w = M$, the fraction extracted, m_e, will be $\tfrac{3}{4}$, and the fraction remaining will be $\tfrac{1}{4}$.

Now suppose the ether is used in two portions of 50 cm³ each. Then, after the first extraction, $\dfrac{m_e}{50} = 3 \dfrac{m_w}{100}$ and $m_e + m_w = M$. The fraction extracted will be $\frac{3}{5}$ and the fraction remaining will be $\frac{2}{5}$. By the second extraction, the same fraction namely, $\frac{3}{5}$, of the remaining solute will be removed. This will be $\frac{3}{5} \times \frac{2}{5}$ of the original mass of solute, and the total fraction now extracted will be $\frac{21}{25}$. The fraction remaining will be $\frac{4}{25}$, which is less than $\frac{1}{4}$.

Expt. 5e-2   Distribution coefficient of iodine between water and carbon tetrachloride

Place 25 cm³ of carbon tetrachloride in each of three bottles or conical flasks. Add 0·25 g of iodine to one, 0·5 g to the next and 1 g to the other. When the iodine has dissolved, add 200 cm³ of water to each flask, shake well and leave overnight. Separate the two layers and titrate 100 cm³ of the water layer and 10 cm³ of the carbon tetrachloride layer with M/10-thiosulphate. Go cautiously with the titration of the water layers as there is very little iodine in them and only enough for two titrations. Calculate the ratio of the iodine concentrations in the two layers. This is the required co-efficient and should be the same in the three experiments.

Expt. 5e-3   Investigation of the reaction $KI + I_2 \rightleftharpoons KI_3$

Application of the distribution law enables us to study some otherwise elusive homogeneous equilibria. To find the equilibrium constant for the above reaction, we must know the concentrations of all three reactions at equilibrium. This may be done by dissolving iodine in carbon tetrachloride and water as above, adding a known weight of potassium iodide to the water layer (it is insoluble in carbon tetrachloride), and titrating the iodine in the two layers. The free iodine in the water layer is in equilibrium with the iodine in the carbon tetrachloride layer, hence its concentration can be found by titrating the iodine in the carbon tetrachloride layer and using the value of the distribution coefficient found above. The titration of the water layer gives the total concentration of iodine, both as I_2 and as KI_3. The concentration of the latter is then found by subtraction. The weight of potassium iodide added originally is known, so another subtraction gives the equilibrium concentration of the KI. All concentrations should be expressed as moles per litre.

Place 25 cm³ of carbon tetrachloride in each of three bottles or conical flasks. To one add about 0·25 g of iodine, to the second and third add about 0·5 g and 1 g of iodine respectively. When the iodine has dissolved, add 100 cm³ of water containing 1 g of potassium iodide to each of the three flasks. Shake well and leave overnight. Separate the layers by means of a separating funnel and titrate the iodine with M/10-thiosulphate, using 50 cm³ of the aqueous layer and 10 cm³ of the carbon tetrachloride layer. Calculate the equilibrium constant in the three cases in the following manner.

Let the concentration of the potassium iodide in the water as both KI and KI³, calculated from the weight of iodide used, be a moles per litre. Let the concentration of the iodine in the carbon tetrachloride layer, from titration, be b moles per litre, and the concentration of iodine in the potassium iodide solution, as both I_2 and KI_3, from titration, be c moles per litre. Let the distribution coefficient of iodine between carbon tetrachloride and water be D. Then, for the equilibrium in the aqueous phase, we have

$$\frac{[I_2] [KI]}{[KI^3]} = \text{a constant.}$$

The concentration of the iodine as I_2 in the water is Db. Hence, the concentration of iodine as KI_3 is $(c - Db)$. The concentration of potassium iodide is therefore $[a - (c - Db)]$. Insert these values in the equilibrium law equation and calculate the equilibrium constant.

Expt. 5e-4 **The distribution of acetic acid between water and ether**

Place 25 cm³ of ether in each of three 100 cm³ conical flasks fitted with bungs. To the first add 25 cm³ of 2M-acetic acid, to the second add 25 cm³ of M-acetic acid, and to the third add 25 cm³ of M/2-acetic acid. Insert the bungs and shake the flasks well for about ten minutes. By means of a separating funnel run the ether and water layers from the first flask into separate flasks. Insert a bung in the flask containing the ether to stop evaporation. Titrate 10 cm³ portions from each layer with M/2-sodium hydroxide solution, using phenolphthalein as indicator. Before titrating the ether layer add an equal volume of water and shake well during the titration. Repeat the titration with each layer from the other two flasks.

Calculate the concentrations (in moles per litre) of acetic acid in the water layer (c_1) and in the ether (c_2). Plot graphs of (i) c_1

against c_2, and (ii) c_1 against $\sqrt{(c_2)}$. Acetic acid is almost un-dissociated in water. What conclusion can you draw about its state when dissolved in ether?

Partition chromatography

The difference in the distribution coefficients of two substances may also be used to separate them. Using the methods of partition chromatography, the constituents of a mixture of a great variety of substances may be separated. Some examples are described below.

In paper chromatography, the substances are separated by using differences in the manner in which they are eluted or moved by a solvent through filter paper or paper specially prepared for chromatography (Expt. 5e-5).

In gas-liquid chromatography, the substances are transported in the vapour phase by a carrier gas through a fixed phase supporting a non-volatile liquid in which the substances dissolve to differing extents (Expt. 5e-6).

Expt. 5e-5   Paper chromatography

(1) Separation of plant pigments.

Grind up a handful of grass in a mortar with not more than 5 cm³ of alcohol or acetone until a deep green extract is obtained. This contains chlorophyll, carotenoids and xanthophyll. Place a few drops in the centre of a filter paper supported horizontally on an empty beaker, and allow them to spread out. Carefully add a few drops of solvent to the centre of the paper and notice that at least two concentric rings are formed, one green and the other yellow. A better separation may be obtained by eluting with a different solvent, e.g. petroleum ether or toluene.

The separation occurs because the coloured substances are differently distributed between the solvent or elutriant and the water of the filter paper fibres.

Repeat the experiment with extracts from coloured leaves and flowers.

(2) Detection of amino-acids.

(i) Samples of the following amino-acids will be needed: glycine, alanine, valine, leucine, proline, cystine. Make separate solutions of the amino-acids by dissolving about 0·1 g in 10 cm³ of water containing 2 drops of concentrated hydrochloric acid. As elutriant use a mixture of 80% propanol and 20% water, or a freshly prepared solution containing 80% pure phenol and 20% water. Cut some strips of filter paper about 1 cm wide and 10 cm long so that each will slip into a test-tube.

By means of a wire or fine glass point, place a drop of one of the amino-acid solutions 1 cm from the bottom of a paper strip. The drop should be no more than 2 mm in diameter. Use a teat pipette to put about 2 cm³ of the elutriant in a dry test-tube without wetting the sides. Drop in the paper strip and cork the tube. The spot of amino-acid solution must be above the level of the liquid in the test-tube. Leave the tube undisturbed until the liquid nearly reaches the top of the paper. This may take over an hour. Remove the paper, holding it with forceps, and spray it evenly with a solution containing 1% ninhydrin in ethanol, using a scent spray for the purpose. Gently warm the paper and look for the appearance of a purplish spot. Outline it in pencil to mark its position.

Repeat with the other amino-acids and compare the positions and colours of the spots.

(ii) Make a mixture of two, or three, amino-acid solutions and find out to what extent this method enables you to separate and detect them.

(iii) Try to break down a protein into amino-acids to find out which are formed. Add about 0·1 g of a protein-containing material (e.g. gelatine, hair, powdered milk, egg albumin) to 5 cm³ of hydrochloric acid made by diluting 30 cm³ of concentrated acid with 2 cm³ of water. Place this in a small flask with a side-arm and fitted with a 'cold finger' to act as a reflux. Add a small piece of porous pot and boil the liquid for an hour. Remove the 'cold finger' and substitute a rubber stopper with capillary tube. Connect the side-arm to a water-pump and evaporate the contents of the flask to a syrup. Stir in 1 cm³ of water and evaporate again under reduced pressure. Dissolve the syrup in 0·5 cm³ of water to make a solution suitable for chromatography. Use this as described above and compare the position and colour of the spots with those obtained from known acids under the same conditions.

Expt. 5e-6 Gas-liquid chromatography

The distribution principle can also be applied to the separation of gases and vapours. The two phases in which the gases are 'dissolved' are (i) a stationary phase consisting of a non-volatile oil absorbed on an inert solid, (ii) a moving phase, consisting of a gas such as nitrogen, hydrogen, etc.

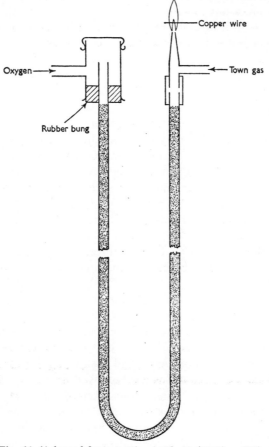

Fig. 66. (Adapted from *J. Chem. Ed.* 40 (1963) p. 539)

A simple apparatus for introducing gas chromatography is illustrated in Figure 66. It consists of a glass U-tube about ¾ m long and 5 mm internal diameter. One end is sealed to a wider tube with a side-arm, the other end is joined to a T-piece.

The stationary phase is prepared as follows. About 10 g of celite (or brick dust), graded to 100–120 mesh, is well mixed in a large evaporating dish with a solution containing about 10 g of silicone oil (M.S. 702) in 100 cm³ of methanol. The methanol is then evaporated off over a water bath. When dry, the granular powder is fed into the U-tube a little at a time until both arms are almost full.

Oxygen from a cylinder will form the moving phase and the flow must be controlled by a needle-valve (e.g. Type OS.IC by Edwards High Vacuum Ltd). The detector consists of a flame of town gas for burning in the oxygen. The substances to be detected will be chlorinated hydrocarbons which, if a copper wire is supported in the flame, will cause the appearance of a green colour. Suitable compounds are ethyl bromide, ethyl iodide, dichlormethane, chloroform, etc.

The mixture is injected into the gas stream from a hyperdermic syringe through a rubber cap ('Subaseal') wired on to the wide tube attached to the column. The needle is pushed through the seal and one drop of the liquid allowed to fall directly on to the celite filling.

To operate the column, first turn on the oxygen gently and adjust the town gas so that, when lit, the flame is 6 cm high. Adjust the oxygen pressure so that the flame is just non-luminous. Start by injecting one drop of a single liquid and, at the same time, start a stop-clock. Note the time when the flame becomes coloured. Wait until the colour goes and again note the time. Repeat with a drop of a different compound and note the times of appearance and disappearance of the coloration in the flame. These times may be of the order of $\frac{1}{2}$ to 2 or 3 min.

Now attempt to detect the constituents present in a mixture of the chlorinated hydrocarbon. Place the mixture in the hyperdermic syringe and again use only one drop. To what extent can measurement of the times of appearance and disappearance of the green colour identify the constituents of the mixture?

6 Solutions

(a) The vapour pressure of solutions

Expt. 6a-1

Set up two barometer tubes as in Expt. 3b-2. Introduce about 1 cm³ of ether into one and about 1 cm³ of a concentrated solution of naphthalene in ether into the other. Note the vapour pressures.

It was known to Faraday in 1822 that the vapour pressure of a solution is lower than that of the pure solvent. A quantitative relation between the concentration of the solution and the lowering in vapour pressure was found by Wüllner in 1856. The subject was studied extensively by Raoult, 1886–90, who was able to formulate the following empirical laws:

(1) The lowering of the vapour pressure relative to the vapour pressure of the solvent is independent of temperature.
(2) The relative lowering of the vapour pressure is proportional to the molar concentration of the solution (at constant temperature).
(3) Equimolecular quantities of different non-volatile solutes in the same quantity of solvent produce the same lowering of vapour pressure (at constant temperature).

The first two laws were based on earlier observations, the third is the most striking and is of far-reaching consequence. It means that the lowering of the vapour pressure of a solvent is proportional only to the *number* of solute particles and is independent of their nature. The laws may be summarized by the expression

$$\frac{p - p'}{p} = \frac{n}{N + n},$$ which approximates to $\frac{n}{N}$ in dilute solutions,

where p is the vapour pressure of the solvent, p' that of the solution, and n is the number of molecules of solute dissolved in N molecules of solvent.

THE DETERMINATION OF MOLECULAR WEIGHTS

This relation may be used to deduce the relative molecular weights of solute and solvent from measurements of the vapour pressures of solvent and solution. If the molecular weight (m) of the solvent is known, that of the solute (M) may then be determined. Suppose that the dissolution of W grams of a non-volatile solute in w grams of solvent of molecular weight m lowers the vapour pressure of the solvent from p to p', then

$$\frac{p - p'}{p} = \frac{W/M}{w/m}$$

whence, if m is known, M may be calculated.

The result gives the molecular weight of the solute *as it exists in the particular solution examined.* An insight into the state of substances in solution can be obtained in this way. In general, substances fall into three classes:

(1) Substances whose molecular weights in solution are equal to their formula weights (see Chap. 1b). Many organic substances such as urea and sugar, when dissolved in water, are of this type.

(2) Substances whose molecular weights in solution are less than their formula weights. Most inorganic solutes dissolved in water are in this class. The explanation must be that, in solution in water, they exist as smaller particles than the 'molecules' corresponding with their empirical formulae (see p. 242).

Suppose that of n solute molecules, a fraction α dissociate into x parts. The total number of dissolved particles will then be

$$(1 - \alpha)n + x\alpha n.$$

If M_0 is the molecular weight of the undissociated solute, and M is the molecular weight calculated from the observed lowering of the vapour pressure of the solvent, then

$$M_0/M = \frac{(1 + (x - 1)\alpha)n}{n}$$

or

$$(x - 1)\alpha = \frac{M_0 - M}{M}.$$

In the special case where $x = 2$, i.e. the solute particles dissociate into two parts,

$$\alpha = \frac{M_0 - M}{M}.$$

(3) Substances whose molecular weights in solution are greater than their formula weights. The phenomenon must be due to 'association' of the solute in solution, so that the freely moving particles, or molecules, consist of aggregates two or more times the size of a particle corresponding to the formula weight (see p. 126). For example, benzoic acid exists as $(C_6H_5COOH)_2$ when dissolved in benzene.

Expt. 6a-2   Vapour pressure of solutions

Draw out a glass capillary of less than 1 mm in thickness. Make a small quantity of a concentrated salt solution deep red with some permanganate. Place a small drop of it on a glass slip and dip the end of the capillary into it. Dip the other end into a drop of water and arrange that the two liquids are separated in the tube by a few millimetres of air. Seal the ends and lay the tube on a glass slide. Mark the ends of each meniscus. Leave for a day and examine again. It will be found that some of the water has distilled on to the solution, showing that the vapour pressure of the salt solution is less than that of water. (See also under 'Osmosis'.)

(b) The boiling point of solutions

Expt. 6b-1

Set up two small flasks containing respectively distilled water and a concentrated calcium chloride solution. Determine the boiling points by means of thermometers immersed in the liquids.

In Fig. 67, curve I is the vapour pressure-temperature graph for a pure liquid. Addition of solute lowers the vapour pressure in accordance with Raoult's law, and curves II and III are the vapour pressure-temperature graphs for two solutions for which the ratio of the number of molecules of solute to number of molecules of solvent are c_1 and c_2 respectively. If these three liquids are boiled under an external pressure p, it is clear that the solutions will have to be heated to higher temperatures than the solvent before their vapour pressures become equal to p, i.e. they have higher boiling points. The lowering of the vapour pressure of a solvent by a solute results directly, therefore, in an elevation of the boiling point.

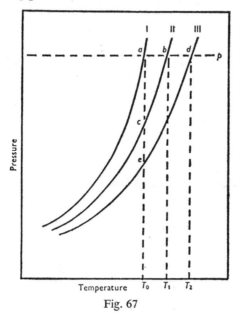

Fig. 67

For dilute solutions, the figures abc, ade approximate to similar triangles, and the elevation of boiling point is proportional to the lowering of the vapour pressure:

$$\frac{ab}{ad} = \frac{ac}{ae}; \qquad \frac{\Delta T_1}{\Delta T_2} = \frac{\Delta p_1}{\Delta p_2}.$$

The latter ratio, by Raoult's law, is equal to c_1/c_2.

Since equimolecular quantities of different solutes in a given quantity of solvent produce the same lowering of the vapour pressure, it follows that they also produce the same elevation of boiling point.

The rise in the boiling point of a solvent produced by dissolving 1 mole of a non-volatile solute in 1000 g of the solvent, is known as the 'molecular elevation constant', K, and the following are the values for a few common solvents (determined from measurements with solutes of known molecular weight):

Water	0·52°C	Ethanol	1·15°C
Acetone	1·70°C	Ether	2·10°C
Benzene	2·70°C	Aniline	3·22°C

If W grams of solute of molecular weight M are dissolved in w grams of solvent, thereby raising the boiling point by $\Delta T°$, then the molecular weight of the solute will be given by the expression

$$M = \frac{K1000W}{w\Delta T}.$$

Measurement of the boiling point of a solution of known composition thus enables the molecular weight of the solute to be determined. It should be emphasized again that the value obtained is for the solute particles as they exist in the particular solution used, and may not (and, in many cases, does not) correspond to the 'formula weight'.

Expt. 6b-2

(i) *Molecular weight of urea by the Landsberger-Walker method*

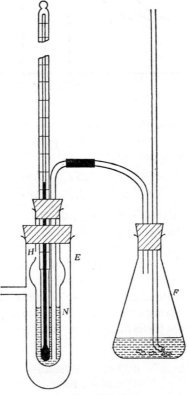

Fig. 68

The liquid is raised to its boiling point in the tube N (Fig. 68) by passing through it vapour from the boiling solvent in F. The tube N has a small hole H near the top, through which the excess vapour escapes into the jacket E, where it maintains the temperature of N. A thermometer reading to tenths of a degree may be used. The results obtained by this method are only accurate to about 5% but the method is quick and superheating of the liquid cannot occur.

Determine the boiling point of distilled water, using about 10 cm³ in N and about 150 cm³ in F. Then introduce a weighed quantity of urea, about 2 g, having first poured away some water, so that there are still only about 10 cm³ in N. Raise the solution to the boiling point, read the thermometer, and at once stop the passage of vapour through the solution. Remove the thermometer from N and read off the volume of the solution. Replace the thermometer, continue to pass in vapour for a few minutes, and take another pair of readings. Given that a mole of solute raises the boiling point of 1000 g of water by 0·54°C in this apparatus, calculate the molecular weight of urea.

(ii) *Elevation of the boiling point of other solvents*

Acetone and alcohol (industrial spirit, 95% ethanol) are good solvents for molecular weight determinations by this method. Their molecular elevation constants are respectively 1·7 and 1·15°C. Benzoic acid or anthracene are suitable solutes. Use about 1·5 g to 10 cm³ of solvent.

(c) The freezing point of solutions

Expt. 6c-1

Dissolve (*a*) 1 g of anthracene, and (*b*) 1 g of naphthalene in separate 10 cm³ samples of benzene in test-tubes. Put some pure benzene in a third tube. Place them in a mixture of ice and salt until crystals appear. Insert thermometers in the tubes and warm them up slowly by the hand, stirring gently. Note the temperatures at which the last crystal in each tube melts.

The phase diagram (Fig. 69) shows the sublimation-pressure curve and vapour-pressure curve for a pure solvent (I). Curves II and III are the vapour-pressure curves of two solutions of molar compositions c_1

and c_2. These curves cut the sublimation-pressure curve at temperatures T_1 and T_2, lower than the temperature T at which the three states of the pure solvent, vapour, liquid and solid, are in equilibrium. The effect of a solute, then, is to lower this temperature. As we have seen, this temperature is called the triple point; it is the freezing point of the substance in equilibrium with its own vapour.

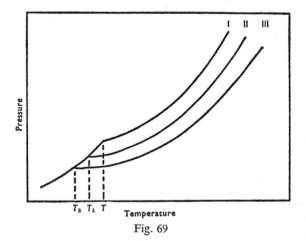

Fig. 69

The effect of pressure on the freezing point is, for most substances, very small, and so under the circumstances in which freezing points are usually measured, viz. atmospheric pressure, the effect of solute on the freezing point is similar to the effect on the triple point. For dilute solutions, a similar relationship holds to that discussed for the elevation of the boiling point, namely, $M = \dfrac{K1000W}{w\Delta T}$, where K is the molecular depression constant for the solvent, and M is the molecular weight of the solute as it exists in the solution under test. One gram-molecular weight of solute dissolved in 1000 grams of solvent depresses the freezing point $K°C$.

The values of K (in °C) for some solvents are as follows:

Water	1·86	Benzene	4·99
Acetic acid	3·88	Phenol	7·50

Measurement of the freezing point of a pure solvent and of a solution of known composition thus provides a method of determining the molecular weight of a solute in the condition in which it exists in the solution measured. It should be noted that the freezing point of a solution of a certain composition is the temperature at which that solution and solid solvent are in equilibrium, and is therefore the temperature at which the smallest quantity of solid separates from the solution on cooling or remains unmelted on warming—*not* the temperature at which the solution solidifies.

Expt. 6c-2 Molecular weight of urea by the lowering of the freezing point of water

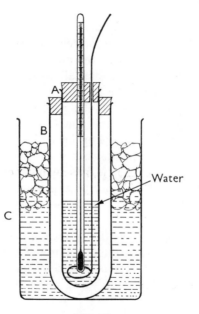

Fig. 70

The apparatus (Fig. 70) consists of a tube *A* carrying a thermometer and surrounded by a wider tube *B*, which acts as an air-jacket. This is immersed in a cooling mixture in the jar *C*. The contents of *A* can be stirred by a glass or copper-wire stirrer. The thermometer, *D*, may be one reading to 0·1 deg C, or may be a Beckmann thermometer, reading to 0·01 deg C.

Determine the freezing point of pure water on the thermometer, using 20 g (20 cm³ from a pipette will do). Cool quickly by immersing the inner tube directly in a freezing mixture of ice, common salt and water at about −5°C. When some ice appears, warm the tube slightly with the hand, place it in its air-jacket and replace in the freezing mixture. Stir continuously. The temperature drops slowly and regularly until freezing begins, when it remains stationary. Some supercooling nearly always occurs, but this should not be allowed to exceed 0·5 deg below the freezing point. If it does, so much latent heat is liberated when the solid separates out that the temperature rises above the freezing point and does not remain constant. In order to induce a supercooled solution to crystallize, stir it vigorously, or, if this fails, add a small piece of the solid solvent.

Next determine the freezing point of the solution of urea. Weigh out from 1 to 1·5 g of urea accurately in a small test-tube, tip the urea into the 20 g of water in the apparatus, and weigh the tube again. Make sure that the urea has all dissolved, and determine the freezing point as before. The freezing point is the temperature at which ice first separates out; as this process continues the solution becomes richer in urea and the temperature gradually falls. It is necessary to prevent supercooling beyond 0·5 deg of the freezing point, for if much ice separates out when the solution freezes, the temperature obtained will be the freezing point of a solution stronger than that made up. Melt the ice by warming the tube in the hand, and repeat the measurement until two concordant readings are obtained.

Add another weighed quantity of urea, about 1 g, and determine the freezing point of this stronger solution. Calculate the molecular weight of urea. The two results obtained should be the same. (One mole of solute in 1000 g of water depresses the freezing point 1·86 deg C.)

Expt. 6c-3

Determine the freezing points of potassium chloride solutions as described above, using first 0·1 g of the salt and making several additions of a similar quantity. Calculate the apparent molecular weight in the different solutions, and compare the results with the formula weight of KCl.

(2) Find the degrees of dissociation of chloracetic acid in aqueous solutions of several compositions by determining their freezing points, using the formula $\alpha = \dfrac{M_0 - M}{M}$ (see p. 136).

Expt. 6c-4   Degree of association of benzoic acid in benzene

Determine the freezing point of benzene as described above using an ordinary thermometer graduated in tenths of a degree. If the benzene is impure, it should be redistilled before use. It freezes at 5·5°C and has a molecular depression constant of 50°C. Many compounds, of which benzoic acid is one, associate in benzene. Determine the freezing point of solutions and calculate the fraction of single molecules that have associated from the formula

$$\alpha = \frac{2(M - M_0)}{M}$$

where M_0 is the molecular weight of C_6H_5COOH, and M is that obtained from the measured depression of the freezing point. Compare Expt. 5e-4 and see p. 92 (hydrogen bonding).

Expt. 6c-5 Molecular weight by Rast's method

Camphor has a very high molecular depression constant; the freezing point is lowered about 40°C by 1 mole of solute dissolved in 1000 g of camphor. It is therefore a useful solvent for the determination of molecular weights of substances that will dissolve in it. As the camphor to be used may not be pure, it is necessary first to measure its depression constant using a solute of known molecular weight. First measure the freezing point of camphor by the capillary tube method (Expt. 4a-2), using a heating bath of medicinal paraffin or tri-cresyl phosphate. Then weigh out exactly about 2·0 g of camphor and about 0·2 g of naphthalene in a small test-tube. Warm it until they just melt, shake the tube in order to mix the contents well and cool rapidly under the tap. Break the tube, grind up the solid, introduce a little into a capillary tube and measure the freezing point by finding the temperature at which the last trace of solid just melts. Repeat the determination with another portion of mixture. The depression produced is so large that a thermometer reading in degrees is sufficiently accurate. From the result and the known molecular weight of the naphthalene, calculate the depression constant of the camphor.

Now use this result to determine the molecular weight of acetanilide, using about 0·2 g of acetanilide in about 2 g of camphor. Measure the depression of the freezing point of the camphor as described above.

Specimen result

Freezing point of camphor	$= 176°C$
Freezing point of a solution containing	
0·247 g of naphthalene in 1·945 g of camphor	$= 145°C$
Freezing point of a solution containing	
0·190 g of acetanilide in 2·089 g of camphor	$=155°C$
Molecular weight of naphthalene	$= 128$

Therefore the molecular weight of acetanilide

$$= 128 \times \frac{31}{21} \times \frac{0·19}{0·247} \times \frac{1·945}{2·089} = 135·3$$

Theoretical value $= 135$

(d) Osmosis

Expt. 6d-1

(i) Place a prune in a beaker of water and another in a strong sugar solution. Examine them after 24 hours.

(ii) Cut two cubes of beetroot or carrot and place one in water and the other in strong salt solution. Examine them after 24 hours and note that the former remains crisp whereas the latter wilts as the fluid drains from the cells.

(iii) Remove the shells from two eggs (stale eggs, unfit for eating, will do) by placing them in strong hydrochloric acid. Wash them and place one in water and the other in strong salt solution. After a day or so, one will be found to have swollen, the other to have shrunk.

In 1748 the Abbé Nollet filled a bladder with alcohol, tied up the neck and placed the bladder in water. The bladder swelled and eventually burst. It is clear that water was entering the bladder faster than alcohol was coming out. Similar processes are occurring in the above experiments; water is entering the prune faster than its juices are coming out, whereas the juices are diffusing out faster than the sugar solution is going in. Hence the prune in water swells, but the prune in sugar solution shrinks. Again, the cell contents of the beetroot or carrot diffuse out faster than the salt solution enters, so the vegetable in salt solution loses its turgidity and wilts. Water enters the egg through the outer skin or membrane faster than water diffuses out, hence the egg in water swells, and may eventually burst.

In all these experiments there is a membrane through which one liquid diffuses faster than another. Such membranes were termed by van't Hoff in 1886 'semi-permeable membranes'.

The examples described above are all of natural membranes; the first semi-permeable membranes to be prepared in the laboratory were made by Traube who published a paper on this in 1867. The most notable was a copper ferrocyanide membrane, which was found to be permeable to water but almost impermeable to many solutes, and has since been one of the most widely used in the study of osmosis.

The process of diffusion of water through a semi-permeable membrane into a solution is called *osmosis*. The statement needs amplifying to make clear that the observed phenomena of osmosis are due, not to a one-way diffusion, but to a difference in the rates of diffusion of the water through the membrane in the two directions. Again, osmosis occurs, not only from pure water into a solution, but from a weaker into a stronger solution.

Expt. 6d-2

(i) Prepare a semi-permeable membrane in a porous pot. A convenient membrane may be made by swilling out the inside of the pot with a strong solution of gelatine containing a few drops of glycerine. Allow to dry overnight and repeat the process. Fill the pot with a

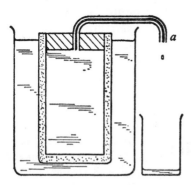

Fig. 71

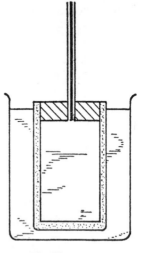

Fig. 72

EBC—L

strong sugar solution or diluted treacle, and fit it with a rubber bung and glass tube as shown in Fig. 71. Immerse the pot in water and examine after some hours.

(ii) Alter the apparatus slightly as in Fig. 72 and examine after a day or two.

An alternative method is to use a piece of visking tubing instead of the porous pot.

In Expt. 6d-1 the semi-permeable membranes form a closed envelope; in those examples where osmosis occurs from the inside of the envelope outwards, the envelope shrinks or crumples, but where osmosis occurs in the other direction, the envelope is forced to swell and a hydrostatic pressure is set up inside it which may ultimately cause it to burst. The arrangement in Fig. 71 enables us to study the building up of this pressure by the process of osmosis. If the apparatus is arranged as in Fig. 71, passage of the water through the membrane into the solution results in the solution overflowing at *a*, and the process of osmosis continues. When the solution becomes, as a result, more dilute, the process will slow down but will continue, in theory, until the solution is infinitely dilute. However, if a slight change is made in the arrangement, as in Fig. 72, so that a hydrostatic pressure is built up as water enters the solution, a new phenomenon is encountered; dilution of the solution does not continue indefinitely but ceases when the hydrostatic pressure reaches a certain value. An equilibrium state is reached where evidently the rates of diffusion of water into and out of the pot have become equal. The establishment of the hydrostatic pressure must in some way be responsible for this, but consideration of the mechanism will be deferred to a later paragraph.

The value of the equilibrium hydrostatic pressure depends upon the particular solution used—its concentration, temperature and the nature of the solute. The first measurements were made by Pfeffer in 1877, using solutions of cane sugar and a semi-permeable membrane of copper ferrocyanide. He found that the equilibrium pressure established was (*a*) approximately proportional to the concentration of the solution (provided this was not too high) at constant temperature, (*b*) approximately proportional to the absolute temperature for a given solution, and (*c*) approximately the same for solutions containing equimolecular concentrations of different substances at the same temperature. Later, measurements were made by Berkeley and Hartley using a slightly different method; instead of allowing osmosis to proceed and to build up a hydrostatic pressure, they applied a pressure to the solution until equilibrium was obtained.

The hydrostatic pressure necessary to prevent the occurrence of osmosis into a solution is called the *osmotic pressure of the solution*. The first two relationships (*a*) and (*b*) mentioned on p. 146 may be expressed in the form $P = kcT$, where P is the osmotic pressure of a solution of molar concentration c at temperature $T°K$. If c is expressed in moles per litre, then in accordance with the third relationship, the constant k will have the same value for all solutes.

This relation holds for a certain class of solute. Van't Hoff pointed out that the observed osmotic pressures of solutions of many solutes were greater than those calculated from the relation $P = kcT$. To obtain the observed value, the calculated value of P had to be multiplied by a factor i, a small number lying, for many solutes, between 1 and 2, $P_{obs.} = iP_{calc.}$. The value of i increases with the dilution of the solution, and, in many cases, approaches a maximum value of 2. These compounds are all binary compounds (e.g. acids and salts), and this behaviour is explained on the hypothesis that the freely moving particle in the solution does not correspond to the 'formula weight' (the 'molecular weight' used in calculating c, the concentration of the solution in moles per litre), but to a smaller weight. The 'formula-weight molecules' of solute evidently dissociate in solution into two parts, the dissociation process being more complete the more dilute the solution, and approaching completion as the solution approaches infinite dilution. The van't Hoff factor i gives a measure of the degree of dissociation. $P_{calc.}$ is proportional to n, the number of 'molecules' of undissociated solute, and $P_{obs.}$ is proportional to the total number of particles after dissociation, namely, $(1 + \alpha)n$, where α is the degree of dissociation. Therefore $i = 1 + \alpha$. This relation holds for binary solutes, but if the solute is dissociated into three parts instead of two, the relation becomes $i = 1 + 2\alpha$.

Using Pfeffer's measurements, van't Hoff showed that the constant k in the relation $P = kcT$ is numerically almost equal to the constant R in the gas equation $PV = RT$. These equations are thus closely analogous and led van't Hoff to state that 'the osmotic pressure of a solution is equal to the gas pressure which the solute molecules would exert if present as a gas in the same volume and at the same temperature as the solution'. This may give rise to the wrong conception of osmotic phenomena; it suggests too close a relation between the 'bombardment pressure' of the solute molecules and the osmotic pressure of the solution—such a relation does exist, but the two pressures are not the same either in nature or magnitude.

The numerical equality of R and k thus remained to be explained. Later (in 1885) van't Hoff deduced thermodynamically a relation between the osmotic pressure and the vapour pressure of a solution, and the gas constant R appeared in the osmotic equation $P = cRT$ by virtue of a step in the proof in which the gas laws are applied *to a gas*. Alternatively, it is possible to show that the numerical equality of R and k results from applying the gas laws and Raoult's law to the osmotic equilibrium, as follows.

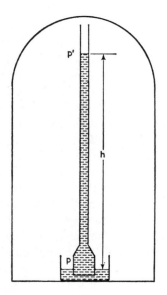

Fig. 73

Consider the arrangement shown in Fig. 73, where a state of equilibrium is depicted between a solution and its solvent across a semipermeable membrane. If p is the vapour pressure of the solvent, p' that of the solution and d_v the density of the vapour of the solvent, then $p - p' = d_v h$. If P is the osmotic pressure of the solution and d_l its density, then $P = d_l h$. Therefore

$$P = (p - p')d_l/d_v. \tag{1}$$

Let there be n molecules of solute to N molecules of solvent. Since by Raoult's law $(p - p')/p = n/N$, equation (1) becomes

$$P = pn/N.d_l d_v. \tag{2}$$

If v is the volume and M the weight of 1 mole of vapour, then $d_v = M/v$, and since $pv = RT$,

$$p = RTd_v/M. \tag{3}$$

The volume of a mass of dilute solution containing N molecules of solvent is NM/d_l, so

$$c = nd_l/NM. \tag{4}$$

Substitute for p/d_v (from (3)) and for nd_l/N (from (4)) in equation (2), and we have

$$P = (cM)(RT/M) \quad \text{or} \quad P = cRT.$$

The mechanism of osmosis

There have been several theories to account for the action of the semi-permeable membrane, but it is unlikely that any one mechanism is adequate to account for all cases. Callendar (1908) assumed that solution and solvent do not meet in the membrane, but that they face each other at opposite ends of numerous minute capillaries through the membrane. Vapour diffuses through these capillaries, distilling from the surface of the solvent on to the solution, above which the vapour pressure is less than above the solvent (see Expt. 6d-6).

The kinetic theory can give a rough picture of the mechanism of osmosis. The rate of bombardment of the membrane by the solvent molecules will be lower on the solution side than on the solvent side of the membrane owing to the obstructive presence of solute molecules on the solution side. The rate of passage of solvent through the membrane is thus greater into the solution than out of it. As the hydrostatic pressure on the solution is increased, the liquid is compressed, and hence the number of solvent molecules bombarding a given area of membrane per second is also increased. As the compressibility of liquids is very small, the pressures needed to cause appreciable effects are great. This picture is open to a number of objections, but at least it helps to explain why the values of osmotic pressures are so high.

Expt. 6d-3   Experiments on osmosis

Cut a slice of beetroot as thin as possible with a razor. Examine it under a microscope with a 25 mm objective, and note that its cells are completely filled with a pink fluid. Remove the slice and cover with powdered sodium chloride. Moisten and leave for 20 minutes. Wash off the salt with a little water and examine again. The contents of the cell are diminished in bulk and shrunk away from the cell wall. Put the slice in water again for about 30 minutes. On examination, the slice should now be found to be quite colourless, as the cells have burst and the cell contents escaped.

Expt. 6d-4

Prepare a solution by mixing 200 cm³ of 10% gelatine, 50 cm³ each of saturated solutions of sodium chloride and potassium ferrocyanide, with 500 cm³ of water. Place the solution in a large jar while still warm and allow to cool. Just before the jelly sets, drop in a few small pills made by grinding 2 parts of copper sulphate with 1 part of sugar and a little water. The pills should be dried off slightly to harden them first. In a few hours curiously shaped growths will appear in the gelatine.

Expt. 6d-5   The chemical garden

Dilute about 20 g of commercial 'water glass' with 100 cm³ of water and place in a wide glass beaker or dish. Cover the bottom with 6 mm of clean sand. Drop in single crystals of a variety of salts, e.g. iron(II) and magnesium sulphates, iron(III), cobalt, nickel and calcium chlorides.

'Growths' of a variety of shape and colour form around the crystals in the course of an hour or so. The skin of insoluble silicate that surrounds each crystal acts as a semi-permeable membrane, and water passes from the weak sodium silicate solution into the stronger salt solution within. The pressure thus produced breaks the silicate skin, which, however, is reformed as a result of the renewed contact between the salt solution and sodium silicate solution, and the cycle is repeated.

Expt. 6d-6   Air as a semi-permeable membrane

Two 6 mm glass tubes are bent at right angles near one end and arranged as in Fig. 74. They are joined by pressure tubing to a

thick-walled capillary tube about 8 cm long. One tube is half filled
with water and the other with strong sugar solution. The two liquids
should fill the rubber connections but be separated in the capillary
by an air-gap about 2 or 3 mm long. Clips on the rubber tubing

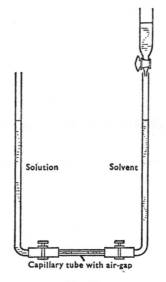

Solution   Solvent

Capillary tube with air-gap

Fig. 74

enable this adjustment to be made. The position of one end of the air-
gap is marked. The clips are closed, and cautiously opened after an
hour or so. The gap moves towards the water, but is brought back
to its original position by adding water to the right-hand tube. Distillation across the gap, with the consequent increase in pressure
on the solution side, continues for many days.

7 Adsorption

Adsorption on solid surfaces

For the following experiments use either 'activated charcoal' or small pieces of wood charcoal which have been heated for some time and cooled in a desiccator.

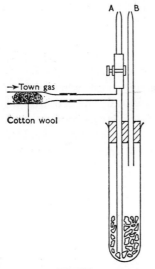

Fig. 75

Expt. 7-1

Invert a test-tube filled with sulphur dioxide or ammonia over a bowl of mercury. Pass a piece of charcoal into the test-tube. The gas is absorbed by the charcoal and the mercury rises, almost filling the test-tube.

Expt. 7-2

Fit up the apparatus shown in Fig. 75. The small tube contains a little cotton wool wet with benzene and the boiling-tube contains activated charcoal. Pass town-gas into the apparatus and ignite it at A and B. Adjust the clip so that the flames are about the same size. Compare the luminosity of the two flames and explain the difference observed. After the gas has been flowing for 2 or 3 minutes, remove the cotton wool and again compare the luminosity of the two flames.

Expt. 7-3

Saturate some water with hydrogen sulphide and shake some of the solution with charcoal. Decant the liquid and compare the smell with that of the original solution.

Expt. 7-4

Prepare a solution of methyl violet about the colour of M/10-permanganate. Shake about 50 cm³ with a teaspoonful of charcoal, and allow to settle. Wash the charcoal until the washings are apparently colourless and then add about 20 cm³ of alcohol. A violet solution is produced. (Show that this is not saturated with the dye by making a more deeply coloured solution in a test-tube.) Decant the solution from the charcoal, add more alcohol and note that a violet solution is again obtained. It is clear that in washing the charcoal, some other factor than the solvent action of the alcohol is at work. This is the adsorptive power of the charcoal, and if sufficient time is allowed, equilibrium is established between the dye in the solution and that adsorbed on the surface of the charcoal.

Expts. 7-1 and 2 illustrate the adsorption of gases by a solid, and Expts. 7-3 and 4 illustrate adsorption of a solute from solution. The importance of the extent and condition of the adsorbing surface may be shown by varying the quantity of charcoal used and by comparing the adsorptive power of the specially activated charcoal with that of stale charcoal taken from the stock bottle. (This matter is discussed in connection with the activity of solid catalysts in Chap. 9c.) Comparisons may also be made with other solids, e.g. crushed roll sulphur, which is much less porous than charcoal. The experiments described below provide further examples of the dependence of the degree of adsorption on the state of the adsorbing surface.

For a given adsorbing surface, the amount of substance adsorbed depends upon its concentration in the solution or gas phase, and upon the temperature. The relation proposed by Freundlich (1906) and known as the Freundlich Adsorption Isotherm, is followed in many instances of adsorption from solution (see Expt. 7-6).

The other experiments illustrate some aspects and applications of adsorption phenomena, including adsorption reagents for use in qualitative analysis, adsorption indicators for certain volumetric analyses, exchange adsorption, and preferential adsorption applied to chromatographic analysis.

Expt. 7-5   Adsorption on freshly formed precipitates

Place about 50 cm³ of dilute permanganate solution in a measuring cylinder and add a few cm³ of dilute caustic soda and a little sodium sulphate solution. Then precipitate barium sulphate by adding a little barium chloride solution. Allow the precipitate to settle, drain off the pink liquid and shake the precipitate with water. Repeat this, and notice that a pink precipitate and colourless liquid are obtained, the permanganate having been adsorbed on the barium sulphate as it passed through the colloidal state before precipitating.

Instead of decanting off the permanganate solution, it may be decolorized by the addition of a little dilute hydrogen peroxide (acidified with dilute sulphuric acid). The precipitate remains pink.

Expt. 7-6   The Freundlich adsorption isotherm

Weigh out six samples of activated wood charcoal or animal charcoal, each exactly 5 g. Place them in conical flasks and shake each with 100 cm³ of oxalic acid solutions of the following concentrations: 0·3, 0·2, 0·1, 0·05, 0·01 and 0·005M. Leave overnight for equilibrium to be established. Filter off the charcoal, and determine the concentrations of the oxalic acid filtrates by titration with standard permanganate or with standard caustic soda, using phenolphthalein as indicator. Also titrate one of the original oxalic solutions.

Specimen result

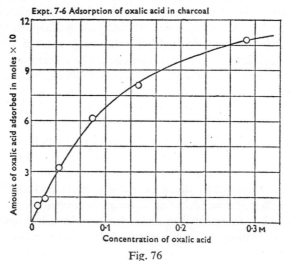

Expt. 7-6 Adsorption of oxalic acid in charcoal

Fig. 76

Calculate the weights of oxalic acid, x (in moles), adsorbed by the charcoal, and plot x against the concentrations, c (in moles per litre), of the solutions with which the adsorbed acid was in equilibrium (i.e. the concentrations of the filtrates, *not* that of the original solutions). Then plot log x against log c. A straight line should be obtained, from the slope of which the index n in the Freundlich adsorption isotherm can be calculated:

$x/m = kc^n$ (m = mass of absorbent, k = a constant).

Expt. 7-7 Adsorption reagents

The adsorption of certain dyes by metallic hydroxides may be used in analysis for the recognition of the metals, and a number of dyes have been discovered to be specific reagents for certain metals. To illustrate their use, dissolve a small pinch of Congo Red in about 100 cm³ of water. Add some aluminium hydroxide, freshly precipitated from about 3 or 4 g of alum with ammonium hydroxide and well washed with water, and bring to the boil. Filter. The dye remains on the hydroxide and the filtrate should be quite colourless.

A very useful reagent is *p*-nitrobenzene-azo-resorcinol (Magneson I), which forms a purple solution and is adsorbed on magnesium hydroxide to give a fine blue lake. Another very sensitive reagent for magnesium is *p*-nitrobenzene-azo-α-naphthol (Magneson II). Use a 0·001% solution in M-alkali, and test for the presence of magnesium either in a very dilute solution or in tap water. The limit of concentration that can be detected is 1 in 260,000.

Expt. 7-8 Adsorption indicators

Certain dyestuffs may be used to indicate the end-point of titrations in which a precipitate is formed. This experiment illustrates the manner in which eosin acts as an indicator in the titration of potassium bromide by silver nitrate.

Put a litre of distilled water in each of two beakers. To each add about 300 g of the sodium salt of eosin. The yellow colour and the green fluorescence are both due to free eosin ions (anions). To one solution add 2 cm³ of M/10-silver nitrate. Add 1 drop of M-potassium bromide solution. The colour deepens to a reddish tint, the fluorescence decreases, and the solution remains transparent. Highly dispersed colloidal silver bromide has been formed, and, as silver ions are in excess, some are adsorbed on the silver bromide to form the complex (AgBr)Ag⁺. Being positively charged, this repels the free silver ions, but attracts the negative eosin ions, causing them to change their colour and tendency to fluoresce. This can be shown by continuing the titration. When a further one or two drops of potassium bromide are added, the red colour first deepens, owing to the formation of more silver bromide; with the addition of about 4 drops of potassium bromide the colour lightens, since the concentration of the excess silver ions and the influence of the adsorption complex are both decreased, thus reducing the adsorption of the eosin. On adding 1 more drop, the equivalence point is reached, and the original colour of the eosin ions returns. The complex is now AgBr with Br$^-$ ions adsorbed, since the eosin ions have been displaced from the surface by the excess of bromide ions. This may be repeated *ad lib* by adding first silver nitrate and then potassium bromide.

TITRATIONS USING ADSORPTION INDICATORS

The table given below gives details of some titrations in which adsorption indicators may be used. Further details may be found in textbooks of volumetric analysis.

To:	Add:	Indicator	Remarks
Chloride or thiocyanate	Silver nitrate	Fluorescein Green fluorescence to pink	In neutral or slightly alkaline solution
Chloride	Silver nitrate	Phenosafranine Red precipitate turns blue	In presence of nitrate ions. Acid solution
Iodide, bromide or thiocyanate	Silver nitrate	Eosin Yellow-red to violet	In acetic acid solution
Ferrocyanide	Lead salt	Alizarin S Yellow to red	Best in M/60 solution

The indicators do not all function in the same way; four possible mechanisms are as follows:
(i) Coagulation of the halogen sol occurs at the equivalence point and at the same time as the adsorption or replacement of the dyestuff ion. Example: the titration of a bromide with silver nitrate, using eosin.
(ii) The sol coagulates appreciably before the equivalence point, and the dye ion passes from the solid state to the solution or vice versa with a change of colour. The indicator is best added just before the equiva-

lence point is reached. Example: the titration of a ferrocyanide with a lead salt, using alizarin S as indicator.

(iii) The precipitate is already well sedimented before the equivalence point is reached. The colour change occurs in the supernatant solution. Example: diphenylamine blue in the titration of a chloride with silver nitrate.

(iv) The colour change occurs on the coagulated precipitate itself. Example: phenosafranine as indicator in the titration of a chloride with silver nitrate.

Expt. 7-9 Exchange adsorption

Shake some powdered wood charcoal with a solution of methyl violet and wash the charcoal with water by decantation. Then add a few cm³ of a weak solution of a detergent such as Teepol, and stir for a few minutes. The water becomes violet in colour owing to the displacement of the adsorbed dye by the Teepol which is preferentially adsorbed.

This experiment illustrates the principle of exchange adsorption or ion exchange. Ion exchange phenomena are of importance in soil chemistry, in water-softening and in many industrial processes. In the soil, metal ions are adsorbed in clay minerals, e.g. montmorillorite, and the 'ion exchange capacity' of a soil is one of its most significant properties. Thus, the addition of lime liberates adsorbed potassium, making it available to plants. Hard water is softened by ion exchange, the calcium and magnesium ions being replaced by sodium ions. The concentration and recovery of uranium from the dilute solutions obtained from its ores is effected by ion exchange. For these processes, a variety of synthetic ion exchange resins are available. See Expt. 7-11.

Expt. 7-10 Preferential adsorption. Chromatographic analysis

(i) Column chromatography

Place a small piece of cotton wool at the bottom of a burette and then fill it with a slurry of alumina (or precipitated chalk) and alcohol. Allow the alcohol to drain out and add a further quantity of slurry until the burette is ¾ full. Drop a small disc of filter paper on to the top of the alumina. Without allowing the column to become dry, add slowly from a pipette a few cm³ of concentrated alcohol extract of grass. Just as the last drop of extract disappears below the filter paper, carefully run in about 10 cm³ of alcohol.

Note the separation of the green band, which is soon washed out of the column, leaving behind the yellow substances, which are more strongly adsorbed on the alumina.

This principle has been used in the separation of a great variety of mixtures, and column chromatography is an important method of separation and purification in many branches of chemistry. For example, it has been used in the separation of plant pigments, the analysis of blood serum, the separation of alkaloids, of pigments and of inorganic cations and anions. Adsorbents such as alumina, magnesia, calcium hydroxide, fuller's earth, etc., may be used in the column. After elution with the solvent, the solutes may be separated into well-defined zones and then removed mechanically, or washed out of the column in succession.

The following examples are also suitable as illustrations of chromatographic analysis on a column:

(a) 5 cm³ of a 0·1% solution of equal parts of methylene blue and malachite green in water. The former is adsorbed as a sharp blue band at the top of the column, whereas the green dye may be completely washed through by distilled water. The blue dye may then be washed through with alcohol.

(b) Examine the differences in the chromatograms of butter and margarine, using 5 cm³ of 40% solutions in benzene.

(ii) *Thin layer chromatography* (*T.L.C.*)

This technique uses the same principle of preferential adsorption, but is faster than the use of a column. In the experiment described below, the adsorbent used is silica gel G, specially prepared and marketed for T.L.C. It is spread uniformly on clean 3 in. × 1 in. microscope slides as follows.

Make a slurry of the silica gel with 2 or 3 times the volume of acetone or chloroform, in a beaker as tall as the slide is long. Hold a glass slide with forceps; immerse it in the slurry and lift it out so that it is uniformly covered with the slurry. Leave it on a flat tile when it will quickly dry.

As an example of T.L.C., solutions of the following dyes may be used: azobenzene, dimethyl yellow, sudan red, brilliant green, indophenol blue, etc. Use a fine wire or glass capillary to spot the test solution on the prepared plate about 1 cm from the bottom. The spot should be no more than 2 mm in diameter. Two or three spots can be placed on each slide. Put the slide with the spots at the lower end in a 250 cm³ beaker containing 4 cm³ of solvent. This consists of a mixture of 70% petroleum ether (b.p. 100–120°C) and 30% benzene. Cover the beaker with a watch-glass and leave it for about 5 to 10 minutes until the solvent nearly reaches the top of the slide. Note the relative distances travelled by the different dyes.

Examine several different colours of ball-point pen inks by T.L.C. Black 'Quink' is suitable and a good separation is obtained using a mixture of 60% isopropanol, 20% ethanol and 20% 2м-ammonia (by volume) as solvent.

Expt. 7-11 *Ion exchange*

(i) Using clay minerals from soil.

First separate the sand, etc., by sedimentation, siphon off the supernatant liquid carrying the clay, precipitate the colloid with ammonium chloride, filter off the clay at the pump, and wash well with dilute hydrochloric acid. Alternatively, start with commercial bentonite or fuller's earth. Use only a paper-thin layer of clay or filtration will be very slow. Pass dilute magnesium sulphate through the filter holding the clay, and wash with distilled water until the filtrate no longer contains magnesium (test with *p*-nitro-benzene-azo-resorcinol-(Magneson I)). Next pass dilute potassium chloride solution through the clay and show the presence of magnesium in the filtrate. The potassium ions have changed places with the ad-sorbed magnesium ions. These ions form part of the crystal structure of the clay minerals, occupying 'exchange positions'.

Exchange may be similarly effected between most metallic ions, e.g. potassium will displace copper and, the reverse, copper will displace potassium. To demonstrate this, pass dilute copper sulphate solution through the clay that has already been treated with potassium chloride, and wash until the washings are free from copper as tested by dilute ammonia. Then pass more potassium chloride through, and show the presence of copper in the filtrate.

(ii) Using synthetic resins.

For these experiments the following two resins will be needed:

'De-acidite FF'. An anion-exchange resin based on polystyrene and containing quaternary ammonium groups.

'Zeo-Karb 225'. A cation-exchange resin based on sulphonated polystyrene.

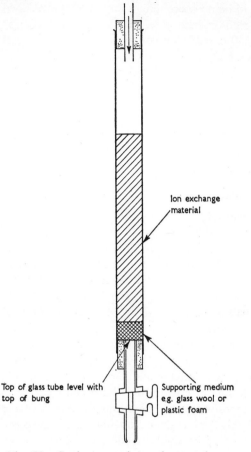

Ion exchange material

Top of glass tube level with top of bung

Supporting medium e.g. glass wool or plastic foam

Fig. 77.   Setting up an ion exchange column

Exchanges are carried out by running the solutions through a column of the resin contained in a vertical glass tube about 30 cm long and fitted with a tap at the lower end (see Fig. 77). A small piece of cotton wool prevents the resin from being washed through the tap.

(1) The removal of copper ions from solution.
Fill the column two-thirds full with Zeo-Karb 225. Pass a little dilute sulphuric acid through the column to make sure that the resin is in the hydrogen form. Wash through with distilled water. Then pour about 250 cm³ of a 5% solution of copper sulphate into the

column. Note what happens. Now again pass dilute sulphuric acid through the column.

(2) The displacement of dichromate by permanganate ions.
Fill the column two-thirds full with De-acidite FF. Run about 250 cm³ of a 5% dichromate solution through the column and wash through with distilled water. Dichromate ions will be held on the resin. Now pass some 1% potassium permanganate solution through the column. What is the colour of the solution that comes out?

(3) The use of a mixed bed to bring about both cation and anion exchange can be illustrated by combining parts of the first two experiments. Use a column containing a mixture of equal parts of Zeo-Karb 225 (in its hydrogen form) and De-acidite FF (converted to its hydroxide form by washing with dilute sodium hydroxide solution followed by distilled water). Through the mixed resin column pass a solution made by mixing equal volumes of 5% copper sulphate and 5% dichromate solutions. What is the colour of the effluent? Pour some dilute sulphuric acid through the column and note the colour of the solution that comes out. Now pour dilute sodium hydroxide solution through the column. What is the colour of the effluent this time?

(4) The use of ion exchange to separate metal ions (nickel and cobalt).
Prepare a 25 cm³ column of De-acidite FF and convert it to the chloride form by passing 25 cm³ of dilute hydrochloric acid through it. Make a solution containing about 4 g of $NiCl_2$ and about 5 g of $CoCl_3$ in 1 litre of 9M-hydrochloric acid. Pour 25 cm³ of this mixed solution into the column. Note that both metal ions are adsorbed, separating slightly into two bands at the top of the bed. Now pass some 9M-hydrochloric acid through the column. The nickel separates as a yellow band and can be washed completely from the column as nickel chloride. Wash the column through with distilled water and note the effect on the adsorbed cobalt. It is washed through as a pink solution.

(For other experiments, see 'Ion Exchange Experiments', available from the Permutit Co. Ltd.)

8 The colloidal state

(a) Colloidal solutions

Expt. 8a-1

To a solution of about 5 g of sodium thiosulphate in a litre of water in a large beaker, add about 10 cm³ of strong hydrochloric acid. Stir, and watch the slow formation of sulphur.

Expt. 8a-2

Pass a convergent beam of light through beakers containing salt solution, sugar solution, arsenious sulphide sol, gold sol, etc. Note the scattering of the light in certain cases, so that the beam is visible as it passes through the solution.

Expt. 8a-3

The following experiment shows the difference in the rates of diffusion of a substance in true solution and colloid. Soak a piece of cellophane in water. Almost fill a small beaker with chlorine water, and place the cellophane over the top. Make a saucer-shaped depression in the cellophane so that the water in the beaker touches the under side (as in Fig. 78). Pour a solution of potassium iodide containing a little starch on to the cellophane. Note that a yellow colour soon appears in the beaker, showing that the iodide has diffused through the cellophane membrane and iodine has been displaced by the chlorine. The starch has not diffused, otherwise a blue colour would have formed. However, the solution above the cellophane turns blue-black, showing that the iodine formed in the beaker has diffused into the starch solution.

In Expt. 8a-1 sulphur is continuously formed as a result of the chemical change occurring. At first the water remains colourless and the sulphur formed is evidently in true solution. Soon an opalescence appears; the solution scatters bluish light and the light that is transmitted is orange-red. If filtered, the solution passes through the filter paper. Eventually a yellow precipitate of sulphur appears and this can be filtered off. As the experiment proceeds, the sulphur particles grow in size—at first they are small enough to be in true solution; finally they are large enough to settle out. In the intermediate stages they are large enough to scatter light, but too small to precipitate. In this condition, they are said to be in the *colloidal state*, and the solution is called *a colloidal solution* or *sol*. A sol is characterized, then, by properties which indicate that the dispersed substance is present in particles of a size intermediate between those in true solution and those which would sediment out. These properties include: (*a*) scattering of light, (*b*) rate of diffusion. It is found, however, that sols are distinguished from true solutions by other properties too, and these will be described later.

Graham (1861) spoke of substances as either 'crystalloid' or 'colloid', according to whether they would diffuse through parchment or not, but it is now known that many substances can exist as both crystalloids and colloids. It is therefore more correct to speak of 'the crystalline state' and 'the colloidal state'. Thus soaps, which to Graham were typical colloids, are shown by X-ray examination to be obtainable in crystalline form, and typically crystalline substances such as common salt and calcite can be obtained as colloids (see Expts. 8c-5 and 6).

A true solution is a homogeneous, one-phase system, but the particles in a colloidal solution are sufficiently large for the system to be regarded as two-phase. The continuous phase (i.e. the water in the case of the sulphur sol) is called the *dispersion medium*, and the colloidal particles are called the *disperse phase*. The properties of the interface between the two phases, i.e. the surface of the particles, are of considerable interest and will be described later.

There are several ways of dispersing a substance in colloidal solution, and examples are described below. The process of freeing a sol from impurities present in true solution makes use of the difference in rates of diffusion through a membrane, and is called *dialysis*. This is described in Expt. 8a-6.

PREPARATION OF COLLOIDAL SOLUTIONS

Note. For all experiments with colloids, great care should be taken to clean all glass vessels used. After treatment with hot acidified dichromate solution they should be well steamed out.

Expt. 8a-4   Sulphur sol

Pass hydrogen sulphide through a wash-bottle of water and then into about 100 cm^3 of distilled water for a minute or so. Pass sulphur dioxide into about 50 cm^3 of water for a minute and add this solution to the other, a little at a time, until there is only a faint smell of hydrogen sulphide. A yellow sol of sulphur is formed. Show that it will pass through filter paper and that it is coagulated by salt solution. Notice the bluish sheen of the sol and the smoky red colours given by transmitted light. After a few days the particles grow in size, and the colour becomes a dull yellow as the sol becomes a suspension.

Expt. 8a-5   Arsenic(III) sulphide sol

Warm about 2 g of arsenic(III) oxide with about 100 cm^3 of distilled water until it has dissolved. Cool, pour off the cold saturated solution, and into half of it pass washed hydrogen sulphide until the solution just smells of the gas. Add some of the arsenic solution until the smell is just removed. A beautiful golden sol is obtained, which will keep in a stoppered bottle for years.

Expt. 8a-6   Iron(III) hydroxide sol

Prepare a strong solution by dissolving a teaspoonful of iron(III) chloride in about three times its volume of water. Filter if necessary. Demonstrate the hydrolysis of the salt by boiling a little in a flask; a brown precipitate appears and the solution is strongly acid. Place about 500 cm^3 of distilled water in each of two flasks and heat one to about 80°C. Pour into both about 3 cm^3 of the strong iron(III) chloride solution; a fine deep reddish brown sol is formed in the hot one. The sol is very stable and will keep for years.

Dialysis of iron(III) hydroxide

Place some of the sol in a trough of cellophane, made by wetting the cellophane and forming it over a wire ring (see Fig. 78). Place this in a basin of distilled water. After a few minutes, test the water for hydrogen and chloride ions. The water remains colourless, showing that the sol has not diffused into it, although the electrolyte has. Change the water every half hour or so. If the dialysis is continued for too long, the sol begins to precipitate, as it owes its stability to the presence of a certain small quantity of hydrogen chloride.

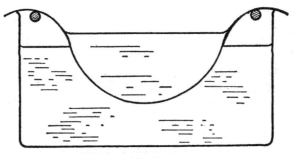

Fig. 78

Expt. 8a-7   Silver sol

A good sol of metallic silver can be obtained by the reduction of silver oxide with dextrin (a hydrolysis product of starch). Dissolve about 5 g of dextrin and 5 g of caustic soda in separate portions of water, mix the solutions and dilute to 250 cm³. Show that this solu-tion will reduce silver nitrate by warming a little with a 10% silver nitrate solution. A silver mirror will be formed in a few minutes. Make a solution of 3·5 g of silver nitrate in 20 cm³ of water and add this slowly to the dextrin solution in a large flask. Warm for 20 minutes on a water-bath; the precipitated silver oxide is reduced to metallic silver, which is in colloidal solution and is black.

To about 50 cm³ of this solution add an equal volume of alcohol. This precipitates the silver. Allow to settle and decant off the liquid after an hour or so. Wash the residue with a little alcohol. Now add 500–1000 cm³ of distilled water. The silver goes into colloidal solution again (is 'peptized'), forming a clear sol that is green by reflected light and brown-red by transmitted light. Add hydro-chloric acid to some of the sol; black silver is precipitated.

Expt. 8a-8

(i) *Faraday's gold sol.* Prepare a few cm³ of a 0·1% solution of gold chloride. Add 1 cm³ to about 200 cm³ of distilled water in a very clean glass bottle. Then add one or two drops of a solution of yellow phosphorus in carbon disulphide and shake well. The colour of the solution changes to red as the gold chloride is reduced to a colloidal solution of gold.

(ii) Gold sols may also be formed by reduction with other reducing agents, for example, tannic acid. Heat a mixture of 1 cm³ of a 0·1% gold chloride solution and 200 cm³ of distilled water to about 60°C. Add 1 cm³ of a freshly prepared solution containing 0·1% of tannic acid and stir well. A red sol is formed. The colour may be deepened by adding further small quantities of the gold chloride and tannic acid solutions.

If the sol is diluted while still hot, the colour changes to purple and blue owing to an increase in particle size. If the reduction is performed by adding 5 cm³ of a 20% solution of tannic acid, the colour of the resulting sol is green.

(b) The stability of sols

Expt. 8b-1

Place some arsenic(III) sulphide sol in a U-tube, insert platinum electrodes in the top of the sol in each limb of the U-tube, and connect to a battery giving about 100 V d.c. Observe the coagulation of the sol around one electrode (which?) and the movement of the yellow sol away from the other electrode.

Expt. 8b-2

Add a few cm³ of a concentrated solution of sodium chloride to (a) some arsenic(III) sulphide sol, and (b) some iron(III) hydroxide sol, and note that the sols soon flocculate. Mix a few cm³ of arsenic(III) sulphide sol and iron(III) hydroxide sol and note that both are precipitated.

Expt. 8b-3

Fill two tall gas jars with water containing clay in colloidal suspension. (Break up a lump of clay, place it in the water and leave it to stand for a week. The coarser particles will settle out, and the remaining suspension is suitable for this experiment.) To one jar, add 50 cm³ of saturated aluminium sulphate solution, and then 50 cm³ of water containing about 5 g of powdered lime. Compare the rate of sedimentation in this jar with that in the untreated jar.

Expt. 8b-4

When an insoluble substance is formed by precipitation in the presence of substances such as gelatine, starch, etc., the rate of growth of the precipitate is retarded, and a sol is first formed. Prepare approximately M/10 solutions of silver nitrate and potassium chromate. Mix equal volumes and note the formation of the insoluble red silver chromate as a precipitate which soon settles. Now add about 50 cm³ of 2% gelatine to 50 cm³ of each solution and mix as before. A red substance forms, but it runs through a filter paper and does not settle out.

FACTORS AFFECTING THE STABILITY OF SOLS

The forces acting on the dispersed particles of a sol, which determine whether it will be stable or precipitate, are:

(i) Gravity, tending to cause precipitation.

(ii) Forces due to the thermal motion of the particles, tending to keep them dispersed.

For particles of colloidal dimensions, the effect of (ii) predominates; thus stable sols, like true solutions, do not settle out. For particles larger than about 10^{-3} mm in diameter, factor (i) is the more important, so if the particles of a sol tend, for some reason, to grow in size, the sol precipitates. The range in diameter of particles in colloidal solution is from about 10^{-3} to 10^{-6} mm; smaller particles than this do not give a 'Tyndall cone' as they are too small to scatter light; they are in true solution. The factors responsible for the stability of a sol are thus those which prevent an increase in the size of the dispersed particles.

In Expt. 8b-1 the particles in the arsenic(III) sulphide sol moved relative to the dispersion medium. This movement of a sol under the influence of an electric field is called *cataphoresis*, and it shows that the sol particles are charged relative to the water; a further study of the phenomenon is made in Expt. 8b-6 below. The potential gradient across the surface of the colloidal particle, from the centre of the particle into the solution, is an important factor in explaining the stability of the sol, for when two particles approach, they repel each other and are thus prevented from coalescing and forming a larger particle. When an electrolyte is added to a sol, the charged particles of the sol adsorb oppositely charged ions on to their surface. The result is that the velocity of cataphoresis is decreased (see Expt. 8b-6) and that the rate of growth of the particles increases, leading to precipitation of the sol (Expt. 8b-2).

The efficacy of the added electrolyte varies with its concentration and also with the valency of its ions (see Expt. 8b-5). A sol whose particles are positively charged is more readily precipitated by trivalent than by monovalent anions; the valency of the cations is immaterial; but a sol with negatively charged particles is more sensitive to multivalent than to monovalent cations, and in this case the valency of the anions is immaterial. This may be explained in a simplified manner as follows: in order to neutralize the charge on a particle, twice as many monovalent ions as divalent ions are necessary. The number of ions adsorbed on the surface of the colloid particle is related to the concentration of ions in the solution by a relation of the form of the Freundlich adsorption isotherm; so the concentration of monovalent ions required to give a certain charge to the sol particles is considerably more than twice the concentration of divalent ions which results in the adsorption of the same charge. This is made clear in Fig. 79.

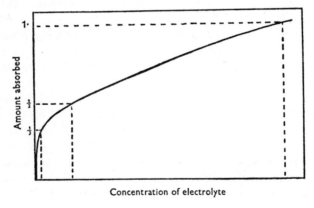

Fig. 79

Origin of the charge on colloidal particles

The charge on certain colloids is due to ionization of the substance itself; thus the particles in a soap sol owe their charge to the loss of sodium ions to the solution. Other colloids become charged by adsorption of ions from solution on to the colloid particles in the course of their formation; sulphur, arsenic(III) sulphide and iron(III) hydroxide sols probably acquire their charge in this way. Once formed, the charge may be modified, and may even change in sign, by changes in the ion content of the dispersion medium.

Expt. 8b-5 Precipitation of sols by electrolytes

Prepare a 2M solution of sodium chloride and put 5 cm³ in the first of a rack of test-tubes. By dilution of the solution prepare M, M/2, M/4, M/8, M/16 and M/32 solutions and put 5 cm³ of each into test-tubes. To each test-tube add 5 cm³ of arsenic(III) sulphide sol, mix and allow to stand for an hour or so. Note the concentration of electrolyte that is just sufficient to precipitate the sol.

Repeat the experiment using solutions of barium chloride of concentrations M/20, M/40, M/80, M/160, M/320, M/640 and M/1280; and again with M/400, M/800, M/1600, M/3200, M/6400, M/12,800, M/25,600 solutions of aluminium sulphate.

Next repeat all three experiments using iron(III) hydroxide sol.

Finally, precipitate the iron(III) hydroxide sol using the following series of solutions: sodium sulphate M/20, M/40, etc., and sodium phosphate M/400, M/800, etc.

Note also the mutual precipitation of the two sols.

Expt. 8b-6

(i) Measurement of the cataphoretic velocity of sols

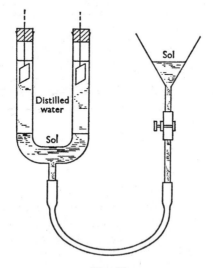

Fig. 80

The arsenic(III) sulphide sol prepared above is suitable for this experiment, or iron(III) hydroxide sol may be used. The apparatus shown is assembled, after washing with hot distilled water. Great care is required in filling the apparatus to get a sharp dividing line between the water and the sol. To do this, pour distilled water into the funnel until the U-tube is one-third full. Close the screw-clip and empty the water from the funnel. Next place the sol in the funnel, making sure that there are no air bubbles in the rubber tube. The level of the sol in the funnel is made somewhat higher than that of the water in the U-tube. Unscrew the clip cautiously so that the sol enters the tube so slowly that the interface remains sharp. When the water reaches the platinum electrodes, close the clip. Connect the electrodes to a d.c. battery of about 120 volts.

Measure the motion of the advancing boundary of the colloid against a paper scale. The cataphoretic velocity should be of the order of 2×10^{-4} cm per sec for a potential gradient of 1 volt per cm.

(ii) *Effect of added electrolyte on cataphoresis*

Refill the apparatus with the arsenic(III) sulphide sol to which dilute sodium chloride has been added so that the concentration of the salt in the resulting solution is about 0·1 g per litre. Note that the cataphoretic velocity is changed. Prepare sols to which increasing concentrations of sodium chloride have been added and measure the cataphoretic velocity of each. Continue to increase the concentrations until the velocity is reduced to zero and the sol precipitates.

Expt. 8b-7 Electric endosmosis

In these experiments high d.c. voltage is required. Great care should be taken not to touch any bare metal components, particularly as, in performing these experiments, the hands may well be wet.

(1) Place copper-foil electrodes in contact with the inner and outer surfaces of a small porous pot containing distilled water, and connect them to a 120 V d.c. battery. The water passes through the clay pot and drips off rapidly if the polarity is in one direction, and the outside of the pot dries out when the polarity is reversed (see Fig. 81).

(2) Make a small 'brick' of wet clay, insert two 15 cm lengths of 4 mm glass tubing, pour distilled water into them and insert copper wire electrodes therein. On connecting them to the d.c. voltage, it

will be found that water rises in one tube and that the clay particles rise into the other.

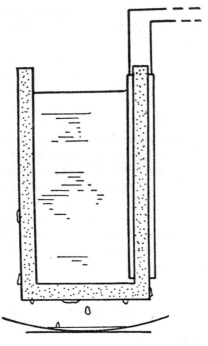

Fig. 81

(3) Put some mud of the consistency of treacle into a beaker. Place two copper-foil electrodes in the mud and connect to the d.c. battery of 120 V. Before switching on the potential difference, notice that, on withdrawing one electrode, the mud adheres to it and only slowly slides off. Then switch on and observe what happens when each electrode in turn is (a) very slowly, (b) suddenly withdrawn. Attempt to explain your observations.

The process illustrated by these experiments is due, like cataphoresis, to the potential gradient across the surface of the colloidal particles into the dispersion medium, the difference being that in electric endosmosis the relative movement of the two phases is manifest as a movement of the dispersion medium rather than of the dispersed particles. The process finds application in the drying of peat, the tanning of leather and in drying walls in a house.

Expt. 8b-8   Smoke precipitation

Smokes are colloidal systems in which the solid particles are dispersed in a gas. Like the dispersed particles of a sol, smokes scatter light, thus often appearing blue in colour. The smoke cloud may include charged particles and this increases the rate of precipitation of the cloud. This phenomenon is used in the electrostatic process for the precipitation of smokes and dusts, which is widely used in industry. The presence of the charge on clouds of fine particles has been responsible for fires and explosions in, for example, coal-mines, flour-mills, etc.

Obtain a glass tube about 40 cm long and 4 cm in diameter, fit it with corks each carrying an exit tube. Push a 60 cm length of copper wire, about s.w.g.16, through the centre of the corks and draw tight. Wind a 2 m length of copper wire round the outside of the tube, fixing the ends with adhesive tape. Mount the tube vertically and fill it with smoke by blowing air through a heated combustion tube containing smouldering brown paper. Attach the wires to an induction coil and note that the smoke rapidly precipitates when the high voltage is applied.

(c) Gels

Expt. 8c-1   Silica gel

Prepare about 1 litre of a solution of sodium silicate by diluting a few teaspoonfuls of water-glass until the specific gravity is about 1·06. Allow to stand, and then decant off the clear liquid. Titrate a portion with M-hydrochloric acid, using phenolphthalein as indicator. Silicic acid is formed and, since the latter is an extremely weak acid, sodium silicate behaves in the titration as if it were caustic soda. Place some of the stock solution in a boiling-tube, add the equivalent amount of hydrochloric acid, and note that the silicic acid formed soon sets to a gel.

Expt. 8c-2   'Solid alcohol'

Prepare a saturated solution of calcium acetate by neutralizing some 50% acetic acid with excess chalk. Filter when cool. Put 10 cm³ of this solution in a beaker, and 90 cm³ of 95% ethanol in another. Mix the two by pouring from one beaker to the other. A gel forms immediately. Cut out a small piece and set fire to it on a wire gauze.

The gel is not stable and separates on keeping. It may be stabilized by the addition of 0·5% stearic acid.

Expt. 8c-3 Rubber

Cut two strips of natural rubber sheet about 5 by 1 cm. Place one in benzene for about an hour, and note that it swells considerably. Leave it in the air overnight and note that it returns to its original dimensions.

Ordinary gelatine provides an excellent example of a typical gel-forming substance. When placed in water, gelatine swells up, absorbing many times its own volume of water. This process of 'imbition' is accelerated by heat. If heated above a certain temperature, the jelly 'melts', owing to the transformation of the gel into a sol. On cooling this solution of gelatine, the viscosity increases, and eventually the whole sets again to a jelly. A gel of this kind is called 'reversible'. The silica gel prepared above is 'irreversible', for once the water has been driven off by heat, the silica will not imbibe water again to form a gel a second time.

A gel may be regarded as a colloidal system in which the dispersed particles of solid differ from those in a sol by being linked up to form a mesh or network. Gels are, in general, formed by substances which have, or can form, long-chain molecules, although other substances can also be obtained as gels (see Expts. 8c-5 and 6).

Gels possess properties usually associated with both the liquid and the solid state; like a solid, a jelly possesses elasticity, but like a viscous liquid, it will slowly flow and take up the shape of the containing vessel.

Natural rubber is another example of a substance that will form a reversible gel. If placed in benzene, it will swell greatly as it takes up the liquid. On leaving in the air, it loses solvent by evaporation and shrinks again.

The preparation and properties of gels are illustrated further by the following experiments.

Expt. 8c-4 Silica gel

Prepare some silicic acid gel as described in Expt. 8c-1. Heat it in a steam oven. It will lose water and harden and may be powdered up in a mortar. This partly dehydrated silica gel is widely used in industry as a dehydrating agent and in making catalysts. It is used as a holder for the catalyst in the contact process for sulphuric acid.

The gel is unaffected by sulphuric acid, withstands continuous high temperature, and is inert to arsenic poisoning. It is used for holding the phosphoric acid in the hydration of ethylene to form ethanol. It is also widely used in adsorption chromatography and in gas-solid and thin layer chromatography.

Expt. 8c-5   Calcium carbonate gel

Dissolve 44 g of anhydrous calcium chloride in distilled water, filter and make up to 100 cm³. Dissolve 57 g of sodium carbonate crystals in distilled water and make up to 100 cm³. Place 10 cm³ of the calcium chloride solution in a boiling-tube and 20 cm³ of the sodium carbonate solution in a pipette. Lower the pipette carefully into the calcium chloride solution and slowly allow the contents to run out, while gradually withdrawing the pipette. In this way, with care, a fairly uniform gel can be obtained. When first formed it is clear, but it rapidly becomes cloudy and, after an hour or so, solid chalk has precipitated.

Expt. 8c-6   Sodium chloride gel

This very beautiful gel is quite easily prepared and is a very good example of the possibility of obtaining a typically crystalline substance in a colloidal state. Thoroughly dry 15 g of sodium salicylate in a hot-air oven and put it into a dry bottle fitted with a glass stopper. In the fume cupboard, add 20 g of thionyl chloride in small quantities. Shake after each addition. Sulphur dioxide is evolved and the reaction takes a few days to complete. A fine greenish-yellow gel forms, and is coloured blue and lilac by reflected light.

Expt. 8c-7   Soap gels in alcohol

Dissolve about 2–3 g of stearic acid in 100 cm³ of 95% ethanol in a conical flask and add a few drops of phenolphthalein. Make an approximately M/10 solution of potash in ethanol by dissolving about 0·5 g of potash in 100 cm³ of the alcohol. Place this in a burette and run into 10 cm³ of the warm stearic acid solution until the phenolphthalein is faint pink. On cooling, the soap formed sets to a good clear gel. It is reversible, dissolving on warming, and reforming on cooling.

Many other gels can be made in a similar manner by neutralizing other fatty acids in hot alcoholic solution with alcoholic solutions of alkali.

Expt. 8c-8   Soap gels in oil

Add about 2% of sodium stearate to 10–20 cm³ of medicinal paraffin oil in a strong test-tube and warm until the soap is dispersed. If any frothing occurs, break the froth by heating the tube around and above the froth. On cooling, a clear gel forms about 200 °C. This becomes opaque below about 140 °C as it breaks down to a paste of micro-crystals of soap in oil. This soap-oil system is a typical *grease*; others may be prepared similarly—lithium and aluminium stearates both give good gels, calcium stearate gels, but separation of oil occurs on cooling ('syneresis'), leaving lumps of gel floating in oil. Sodium salts of other long-chain fatty acids from lauric to cerotic can also be used.

Expt. 8c-9   Peptization of soap gels

(1) A number of substances, when added in small amount to a soap-water mixture, produce a stiff gel. This peptization may be observed by adding to an approximately normal solution of sodium stearate, while hot, enough peptizer to make the resulting solution about 5M. The effect of the following peptizers may be tried: phenol, cresol, aniline, and glycerol and ethanol in larger amounts. On cooling the soap-water mixture, the soap separates as a curd, but where peptization occurs, either a clear solution or a gel is obtained.
(2) Observe the effect on a calcium stearate in paraffin (2–5%) gel of the addition of 2–5% of cresol, or oleic acid. The gel is peptized, and on cooling the molten mixture, it is found that the gelation temperature has been lowered, and that a clear gel results instead of the lumps obtained with no peptizer. A 50 : 50 mixture of cresol and oleic acid is more effective than either alone, addition of less than 1% having a marked effect.

Expt. 8c-10   Rhythmic banding in silica gel

(i) Make up a solution of water-glass of density 1·06. Mix 100 cm³ with an equal volume of M-acetic acid, containing about 3 g of potassium iodide. Pour the solution into tubes to set. Cover the gel that forms with 0·3M-mercury(II) chloride solution. In a few days, bands of red mercury(II) iodide will appear.
(ii) Very sharp bands separated by clear gaps are formed by copper(II) chromate. Make a gel from the 1·06 density water-glass by mixing with an equal volume of 0·5M-acetic acid, and add potassium chromate so that the mixture is 0·05M with respect to it. Allow to gel

in a long tube and cover, when set, with 0·3M-copper(II) sulphate. The bands will form in a few days.

(iii) Add about 2 cm³ of 0·5M lead acetate to 25 cm³ of 0·5N-acetic acid and use this to prepare silica gel from the water-glass solution of density 1·06. Allow the gel to form in a boiling-tube, about half filled. When it has set, cover the gel with approximately 2M-potassium iodide solution. Lead iodide forms on the surface, but after a few days, rings of yellow crystalline lead iodide form in the gel, which recrystallize after some weeks forming small hexagonal plates.

Expt. 8c-11   Liesegang phenomena in gelatine

(1) Dissolve 3 g of gelatine and 0·1 g of potassium dichromate in 100 cm³ of water, and pour into tubes to set. When the gel has set firmly, place some 30% silver nitrate solution on top of the gel and leave undisturbed. Concentric rings of brown silver chromate are formed (see Plate 7b).

(2) Prepare a 3% solution of gelatine containing 5% of magnesium chloride. When set, allow 0·880 ammonia to diffuse into the gel. White rings of magnesium hydroxide will form after several days (see Plate 7a). They are separated by clear gelatine, which can be cut out and shown to be free from magnesium. If a 2% agar gel is used, a spiral of magnesium hydroxide may be obtained.

Expt. 8c-12   Thixotropic gels

Put about 1 cm³ of powdered bentonite or Fuller's earth in a test-tube and add water until the powder becomes a thick mud. Add a few more drops of water until, on shaking the test-tube vigorously, the contents can be heard to be moving inside the tube as a liquid. Now invert the test-tube. No liquid should flow out. Add a little more water and repeat the procedure. The clay and water form a system called a thixotropic gel, which has some rigidity when left undisturbed but flows like a liquid when mechanically agitated. The property is utilized when thixotropic muds are used in drilling, for example, for oil, and in certain paints. Thixotropic paints are sufficiently fluid to be stirred and applied with a brush, but once left undisturbed, do not flow. Various culinary and toilet preparations, such as whipped cream, and shaving lather, are also thixotropic.

(d) Emulsions

Expt. 8d-1

Shake about 5 cm³ of paraffin oil with about 50 cm³ of water an emulsion is formed but rapidly separates into the two liquid layers again. Now add a few cm³ of soap solution, shake, allow the foam to settle, and notice that the oil and water emulsion is now stable.

An emulsion is a colloidal system in which both the dispersed phase and the dispersion medium are liquids. The miscibility of different types of liquid is discussed in Chapter 5b, to which reference should be made. In general, it is found that substances with molecules of similar type, e.g. water and ethyl alcohol, benzene and aniline, paraffin oil and petroleum ether, are miscible with each other in the liquid state, whereas such pairs of liquids as water and benzene or water and paraffin are not. The above experiment shows, however, that paraffin oil may be temporarily dispersed in water as fine droplets by vigorous shaking. The mixture thus formed is not stable, but the addition of a little soap enables a stable emulsion to form. The soap is one example of an 'emulsifying agent'. The soaps, with molecules consisting of long carbon chains terminated by polar groups (metal salts of carboxylic acids), are 'amphiphathic'—the carbon-chain end being soluble in paraffin oil and the polar end being soluble in water. The soap thus acts as an emulsifier by being adsorbed at the oil-water interface, blanketing the oil drops and preventing them from coalescing. Other detergents, such as 'Teepol', consist of sulphonated hydrocarbons and work in a similar manner.

The emulsion produced in this way is an 'oil-in-water' type. The phases may be inverted and water-in-oil emulsions produced by the use of different emulsifying agents (see Expts. 8d-2 and 3).

Emulsions may be 'broken' by the methods employed in precipitating sols, i.e. by addition of electrolytes, or by cataphoresis, or by methods devised to remove the emulsifying agent.

Expt. 8d-2 Emulsifying agents

Attempt to disperse about 0·05 cm³ of paraffin oil (kerosene) in 10 cm³ of each of the following by vigorous shaking and then by 'homogenizing' in a domestic mixer: (*a*) distilled water, (*b*) 1% caustic soda solution, (*c*) 1% sodium oleate solution, (*d*) 1% gelatine, (*e*) 1% bentonite (fine clay) in water, (*f*) 1% iron(III) oxide in water.

EBC—N

Dilute 10 cm³ of each emulsion with 10 cm³ of water in a test-tube. Remove 1 cm³ and add to 1 cm³ of 10% gelatine solution (warm). Place a drop on a microscope slide, cover with a cover-glass, and compare the sizes of the oil drops.

Expt. 8d-3   Emulsion types

Homogenize 50 cm³ of kerosene or benzene with 50 cm³ of 1% sodium oleate solution. Divide into two equal parts and add to one a small quantity of magnesium sulphate. Determine with each part, whether the emulsion is oil-in-water or water-in-oil as follows:
(1) Add a few drops of the emulsion to water and to oil.
(2) Add to small quantities of the emulsion (*a*) an oil-soluble dye (e.g. 'Oil Red'), (*b*) a water-soluble dye (e.g. methylene blue).

The phase inversion brought about by the magnesium sulphate is well shown by means of the oil-soluble dye. In the oil-in-water emulsion, only a few spots will be dyed, but when the emulsion inverts, the whole liquid appears coloured.

9 Rates of reaction

(a) The effect of concentration on rate of reaction

Expt. 9a-1

Make three solutions of sodium thiosulphate of different concentrations by dissolving about 1, 2, and 3 g respectively in equal volumes (300 cm³) of water in three beakers. To each add 1 cm³ of concentrated hydrochloric acid and mix by stirring. Note the times of the appearance of the sulphur.

The rate of a given reaction is influenced by a number of factors: (1) the concentration of the reactants, (2) the temperature, and (3) the presence of the catalysts. These factors will be considered in the following sections.

The above experiment shows that the concentration of one reactant greatly effects the rate of the reaction. The effect of concentration on rate is described by the *Law of Mass Action* which states that 'the rate of a reaction occurring at constant temperature is proportional to the product of the 'active masses' of the reacting substances'. The meaning of the term "active mass", introduced originally in the nineteenth century, depends upon the nature of the reaction. If the reaction is between gases, the 'active masses' of the gases are measured by their partial pressures; if between substances in dilute solution, by their concentrations; and if between miscible liquids, by their molar proportions. The law only applies to reactions occurring in one phase.

Consider the reaction between two substances A and B. Let $[A]$ and $[B]$ represent their 'active masses'. As A and B become converted by the reaction into other substances, $[A]$ and $[B]$ decrease and the reaction consequently slows down. It used to be thought that the rate of the reaction, v, was given by $v = k[A][B]$. There are some reactions which follow this simple rate law, but most reactions are complex and their rates are not directly proportional to their active masses but to some function of them (see p. 188). By measuring the extent of the reaction after various intervals of time, it is possible to calculate the rates at which the reaction is proceeding.

The extent of the reaction is usually gauged by measuring the concentration of either one reactant or one product. These measurements may then be used to test the law of mass action, by using the integrated form of the above equation, which is obtained in the following manner.

Consider the reaction $A + B \rightarrow \ldots$ Let a moles/litre be the initial concentrations of A and of B. Suppose that after time t, x moles have reacted. The concentrations of A and B will then each be $(a - x)$ moles/ litre. According to the law of mass action in its simplest form, the rate of the reaction, i.e. the rate of increase of x with time, dx/dt, is proportional to $(a - x)(a - x)$, i.e. $dx/dt = k(a - x)^2$.

On integration this becomes $kt = - \dfrac{1}{a - x} +$ a constant.

At the start of the reaction, $t = 0$ and $x = 0$, so the constant of integration is $1/a$. Therefore

$$k = \frac{1}{t} \frac{x}{a(a - x)}.$$

This equation may be tested for a given reaction by calculating k for various values of x and t (obtained at constant temperature), or by plotting the function $x/(a - x)$ against t. If a straight line is obtained, the reaction follows this equation and is said to be a 'second-order' reaction, because the rate depends on changes in the concentrations of *two* reactants.

Measurements of the rate may indicate that a reaction does not follow the second order law. For example, if the rate of reaction depends on the change in concentration of only one reactant, the reaction is said to be of the first order and would follow the first order equation dx/dt $= k(a - x)$. On integration, this gives $k = \dfrac{1}{t} \log_e \dfrac{a}{a - x}.$

Reactions of the third or higher order are rare.

THE KINETIC THEORY AND RATES OF REACTION

It is reasonable to assume that the rate of reaction between two substances depends in some way upon the rate of collision between their molecules. The number of collisions occurring per second between molecules of A and B will be proportional to the concentrations of both A and B, so the above assumption is in accordance with the law of mass action. It is difficult to get a kinetic picture of first-order reactions, in which a fixed proportion of the remaining reactant molecules react each second, but a more detailed picture of the mechanism of reactions in terms of the kinetic theory becomes possible as a result of the study of the effect of temperature on reaction rates (see Chap. 9b).

Expt. 9a-2   The rates of decomposition of sodium thiosulphate solutions by dilute nitric acid

The time taken for the precipitate of sulphur to appear is taken as a measure of the rate of the reaction.

Prepare 50 cm³ of approximately $M/2$-nitric acid and about 25 cm³ of approximately 2M-sodium thiosulphate solution. From the latter prepare by dilution 5 cm³ each of solutions of concentrations 3M/2,

Specimen result

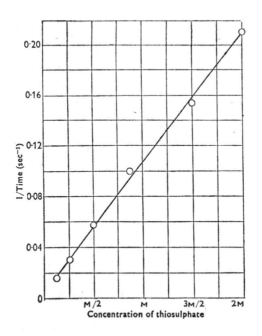

Fig. 82

M, M/2, M/4 and M/8. To the first of these add 5 cm³ of the acid, mix well and simultaneously start a stop-clock. Note the time when sulphur first appears. Repeat with the other thiosulphate solutions. At a temperature of about 18°C the times will range from about 5 seconds for the most concentrated solution to about 1 minute for the weakest.

Plot $1/t$ (as a measure of the rate) against the concentration of the thiosulphate. Since the concentration of the other reactant was the same in each case, a straight line should be obtained, showing that the rate of the reaction is proportional to the concentration of the thiosulphate (see specimen result, Fig. 82).

Expt. 9a-3   *The velocity of saponification of methyl acetate*

The hydrolysis of an ester such as methyl acetate takes place much faster in the presence of caustic soda than in the presence of an acid and, in alkaline solution, is irreversible. It is therefore possible to follow the rate of hydrolysis by alkali by stopping the reaction after a certain time by the addition of a known amount of standard acid. By back-titrating the excess acid, the amount of caustic soda remaining unconverted to sodium acetate can be determined, and hence the extent of the esterification can be measured.

Add 50 cm³ of an approximately M/40 solution of methyl acetate in water to 50 cm³ (from a burette) of exactly M/20-caustic soda. Cork up the mixture in a conical flask and place in a water-bath to keep it at a constant temperature, say 15°C. Every 5 minutes withdraw 10 cm³ with a pipette, add to 10 cm³ of exactly M/20-hydrochloric acid and note the time. Find the amount of acid left by titration with M/20-caustic soda, using methyl orange as indicator:

$$CH_3.COO.CH_3 + NaOH \rightarrow CH_3.OH + CH_3.COONa$$

Concentration after time t (moles/litre):

$$(a - x) \qquad (b - x) \qquad x \qquad x$$

The velocity of the reaction is given by

$$dx/dt = k(a - x)(b - x),$$

which on integration gives

$$k = \frac{1}{t} \frac{1}{(a - b)} \log_e \frac{b(a - n)}{a(b - x)}.$$

By calculating the values of $\log_{10} \dfrac{(a - x)}{(b - x)}$ and plotting them against values of t, a straight line should be obtained, showing that the reaction follows the second-order equation. The results are worked out as follows: From the volume of alkali used to neutralize the excess acid, calculate the number of cm³ of M/20-alkali remaining in 10 cm³ of reaction mixture after time t. Convert this to moles per litre to get $(b - x)$. The initial value b is known, hence find corresponding values of x. The sodium hydroxide in the original mixture was in excess of the ester, so that the final value of x is equal to a, the initial concentration of the ester. Knowing this value, calculate the values of $(a - x)$. It is thus necessary to allow the reaction to continue until all the ester is saponified; at 15°C, this will take at least 1 hour.

Specimen result

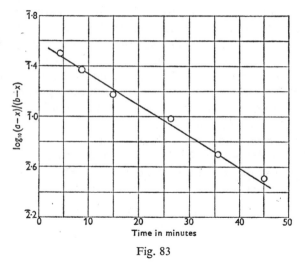

Fig. 83

Expt. 9a-4   The reaction between hydrogen peroxide and hydriodic acid, $H_2O_2 + 2HI \rightarrow 2H_2O + I_2$

Hydrogen peroxide liberates iodine from acidified potassium iodide at a rate which depends on the concentrations of the peroxide, the iodide and the acid. If the reaction (which is irreversible) is carried out in the presence of sodium thiosulphate, the iodine is converted back to hydriodic acid as fast as it is formed, and the concentration of the latter therefore remains unchanged. In any given 'run' the rate of the reaction is therefore dependent only on the decreasing hydrogen peroxide concentration and the first-order law is followed. However, in solutions of higher acidity or iodide concentration, the velocity constant is greater. It is also greater at higher temperatures, so the temperature should be kept constant during the experiment.

First find the strength of the hydrogen peroxide to be used, expressed in terms of its equivalent of thiosulphate: dilute 10 cm³ of fresh '10 volume' hydrogen peroxide (about 2M) to 100 cm³ and titrate with standard permanganate. Then determine the concentration of the stock 'M/10'-thiosulphate in terms of the permanganate by adding acidified potassium iodide to 25 cm³ of the permanganate and titrating with the thiosulphate in the usual way.

Fill a burette with the thiosulphate solution. Put 800 cm³ of distilled water in a 1 litre flask, add 40 cm³ of approximately M-sulphuric acid and a little starch solution. Dissolve 4 g of potassium iodide in a little water and add to the litre flask just before starting the reaction. Run in 2 cm³ of thiosulphate from the burette, and start the reaction by adding 10 cm³ of the undiluted '10 volume' peroxide from a pipette. Start a stop-clock when most of the peroxide has run in. When the iodine produced by the reaction is in excess of the added thiosulphate, a blue colour will suddenly appear; note the time of its appearance and run in another cm³ of thiosulphate. Note the time when the blue colour reappears, and continue in this manner until about a dozen readings have been taken.

Let a moles per litre be the initial concentration of the peroxide and $(a - x)$ the concentration after time t. Then, according to the law of mass action,

$$dx/dt = k(a - x),$$

whence

$$k = \frac{1}{t} \log_e \frac{a}{a - x}.$$

The number of cm³ of thiosulphate, n, added after time t is a measure of x, the amount of peroxide decomposed. Suppose m was the number of cm³ of thiosulphate equivalent to the initial quantity of peroxide, then m is proportional to a and $(m - n)$ is proportional to $(a - x)$. Calculate the values of $(m - n)$ corresponding to the measured values of t, and plot a graph of t against $\log \dfrac{m}{(m - n)}$.

This will be found to be a straight line, showing that the reaction is of the first order. See specimen results (Fig. 84).

If time allows, repeat the experiment (i) using a different concentration of potassium iodide, say, 10 g instead of 4; (ii) at a temperature 10 °C higher (heat the 800 cm³ of water to about 25 °C and
place the flask in a large bucket of water kept at about this temperature); (iii) in the presence of ammonium molybdate as a catalyst: add 1 cm³ of a solution containing 0·1 g of molybdate in 100
cm³ of water. Show that a first-order constant is obtained, but that
the value of the constant is in each case greater than that obtained
in the original experiment.

Specimen results

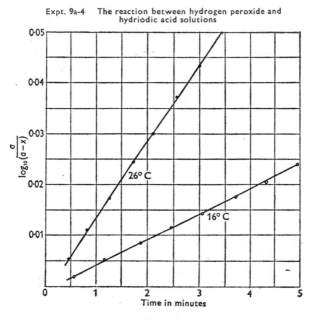

Expt. 9a-4   The reaction between hydrogen peroxide and
hydriodic acid solutions

Fig. 84

(b) Temperature coefficient of reaction rates

Expt. 9b-1

Dissolve about 1 g of sodium thiosulphate in about 150 cm³ of
water and place 50 cm³ of the solution in each of three beakers.
Leave one at room temperature, heat one to about 30 °C and the

third to about 60°C. Add 1 cm³ of concentrated hydrochloric acid to each simultaneously. Mix well by stirring and note the times of appearance of the sulphur.

Under comparable conditions, all chemical reactions proceed faster at higher temperatures. The speed of a reaction is conveniently measured by the velocity constant (see Chap. 9a), as this is independent of the concentrations of the reactants. The temperature coefficient of the velocity constant varies from one reaction to another; it is usually about doubled for a rise in temperature of 10°C. The variation of velocity constants with the absolute temperature was found to be expressed by the empirical relation: $\log k = A - B/T$, where A and B are constants.

EXPLANATION IN TERMS OF THE KINETIC THEORY

Measurements of the rates of chemical reactions, in terms of the number of moles reacting per second, show that only a fraction of the molecules which collide undergo reaction. It might be assumed that the rate of reaction was proportional to the number of molecular collisions, but this simple view will not account for the observed variation of reaction velocity with temperature according to the relation: $\log k = A - B/T$.

In 1889 Arrhenius put forward a theory to account for the observed temperature coefficient of reaction rates. He suggested that only certain 'active molecules' in the gas react on collision, and that these active molecules differ from the rest in possessing exceptionally high energies. The critical value of the energy necessary for reaction to occur is called the *energy of activation*. The nature of this excess energy was not specified, but it might be in the form of kinetic energy due to the high velocity of the molecule, or in the form of energy due to rotation of the molecule, or in the form of vibrational energy of the atoms constituting the molecule, etc. According to the kinetic theory of gases, a random distribution of kinetic energy is maintained among the molecules by molecular collisions, and Clerk Maxwell showed that the fraction of the total number of molecules which, at any moment, possess energy greater than any assigned value E, is $e^{-E/RT}$ (see p. 40). Thus, $n = n_0 e^{-E/RT}$, where n_0 is the total number of molecules and n is the number with kinetic energy greater than E. If the 'active molecules' of Arrhenius are assumed to be merely those with kinetic energy greater than E, the energy of activation, it follows that the rate of reaction is proportional to n. Hence, if the velocity constant is k,

$$k = Cn_0 e^{-E/RT}$$

where C is a constant. Taking logarithms, we have

$$\log_e k = \text{a constant} - E/RT,$$

an equation of the same form as the empirical relation between velocity constant and temperature.

Thus the theory of Arrhenius is consistent with the manner in which reaction rates are found to vary with temperature. By measuring k at various temperatures, and plotting $\log k$ against $1/T$, a straight line is obtained, from the slope of which the value of E, the activation energy, may be calculated (slope $= -E/R$). (See Fig. 85.)

In a given reaction, the total number of collisions per second (n_0) may be calculated from the kinetic theory, and using the value of E obtained from the temperature coefficient of the rate, the number of collisions which should be effective in producing reaction (n) may be obtained. When this value is compared with the measured rate of the reaction it is found that most reactions proceed faster than indicated by this simple theory. This means that the source of the activation energy is not only the molecular kinetic energy, in fact, increased velocity of the molecules does not greatly increase the rate of reaction, and energy stored up in the molecule in other forms contributes towards the energy of activation. In the case of reactions occurring at surfaces, the absorbed reactant molecules receive energy from the solid surface and the reaction may proceed many hundreds of times faster than the same reaction occurring in one phase (see p. 188).

Examples

(1) *Decomposition of hydrogen iodide.* From the temperature coefficient of the rate of decomposition, the energy of activation, E, is found to be 43·7 kcal/mole. At a temperature of 283 °C and a concentration of 1 mole/litre, the number of collisions per second, n_0, occurring between the molecules contained in 1 cm^3 (as calculated from the effective diameter of the molecules, and their mean velocity) is $6·7 \times 10^{31}$. The number of collisions between activated molecules is given by $n = n_0 e^{-E/RT}$, and is $3·3 \times 10^{14}$ per sec in 1 cm^3 (or $3·3 \times 10^{17}$ in 1 litre). Expressed as the fraction of 1 mole reacting in 1 second, this becomes $5·3 \times 10^{-7}$. The experimentally measured value of the velocity

constant expressed in the same units is $3 \cdot 5 \times 10^{-7}$, a figure of the same order of magnitude as that calculated from the assumptions of the kinetic theory.

(2) *The attack of platinum by iodine vapour.* The rates of this reaction are quoted to illustrate that surface reactions proceed faster than indicated by rate calculations based on the relation $n = n_0 e^{-E/RT}$. From the temperature coefficient of the rate of attack, E is found to be $57 \cdot 8$ kcal/mole. At a temperature of $1180 °C$, with the iodine gas at a pressure of $0 \cdot 01$ mm of mercury, the measured rate of attack of 1 cm² of platinum surface is $1 \cdot 0 \times 10^{-12}$ moles per second. The rate calculated above by assuming a simple collision process to be the means of activation is much lower, namely, $0 \cdot 8 \times 10^{-17}$ moles per second. By assuming that the reacting iodine derives activation energy from the hot platinum surface, the estimated rate of reaction becomes $0 \cdot 25 \times 10^{-12}$ moles per second, which is of the same order as the measured value.

THE MECHANISM OF REACTIONS

The study of the kinetics of a reaction may lead to an understanding of its mechanism. Measurements of the manner in which the rate depends on the concentrations of the reactants enable hypotheses to be made about the way the molecules are reacting. For example, in the reaction between iodine vapour and a hot platinum surface quoted above, it is found that provided the pressure of iodine vapour is low, the rate of production of platinum iodide is first order with respect to the iodine, i.e. the rate decreases according to the first order law as the pressure falls off. This strongly suggests that the rate of reaction is determined by the rate at which the iodine molecules collide with the platinum surface. However, if the iodine pressure is higher than a certain value (depending upon the temperature of the platinum surface), the reaction becomes of zero order, i.e. the rate is independent of the pressure of iodine vapour. This can be interpreted by the following mechanism: the rate of bombardment of the platinum surface is higher than the rate of evaporation of platinum iodide, so the surface is kept supplied with iodine and the rate of formation of platinum iodide is independent of the pressure of iodine vapour. This suggested mechanism is supported by the fact that at higher temperatures of the platinum, the pressure at which the reaction changes from first order to zero order is higher.

The rate of decomposition of hydrogen iodide is a reaction that has been much studied. It is found to follow a second order rate law,

i.e.
$$\frac{d[I_2]}{dt} = k[HI]^2.$$

The mechanism may therefore be that suggested by the equation

$$2HI \longrightarrow H_2 + I_2,$$

i.e. reaction occurs between two hydrogen iodide molecules when they collide. Such a reaction would be described as 'bimolecular'. It must be pointed out that not all reactions which look, from their chemical equations, as though they might be bimolecular, follow a second order law. The term 'molecularity' refers to the mechanism of the reaction (and moreover, often only to one step in the reaction), and the term 'order' refers to the experimentally observed manner in which the rate varies.

A reaction where the rate follows a rather complicated relation is that between hydrogen and bromine:

$$H_2 + Br_2 \longrightarrow 2HBr,$$

namely,
$$\frac{d[HBr]}{dt} = k \frac{[H_2][Br_2]^{\frac{1}{2}}}{1 + \dfrac{[HBr]}{m[Br_2]}}$$

Quite an involved mechanism must operate to give rise to this rate law. The mechanism assumes a chain reaction in which bromine atoms are formed as intermediaries:

$$Br_2 \longrightarrow 2Br$$

These are then assumed to react with hydrogen molecules forming atoms:

$$Br + H_2 \longrightarrow HBr + H$$

The bromine atoms are then reformed by the reaction:

$$H + Br_2 \longrightarrow HBr + Br$$

The chain reactions may be terminated by such reactions as:

$$Br + Br \longrightarrow Br_2, \quad \text{and} \quad H + HBr \longrightarrow H_2 + Br$$

By assuming that a 'stationary state' is set up in which the various particles involved are formed and react at equal rates, the above rate law for the overall reaction can be deduced.

Expt. 9b-2   Temperature coefficient of rate of decomposition of sodium thiosulphate by dilute nitric acid

As in Expt. 9a-2, the time taken for the appearance of the precipitate of sulphur is taken as a measure of the rate of reaction. Prepare 50 cm³ of approximately M/2-nitric acid, and 50 cm³ of approximately M/4-sodium thiosulphate solution. Place 5 cm³ portions of the acid in each of eight test-tubes, and 5 cm³ portions of the thiosulphate in each of another eight tubes. Add a tube of acid to one of thiosulphate, mix well, and simultaneously start a stop-clock. Note the time when the sulphur first appears. Take the temperature of the mixture. Place the remaining tubes in a large beaker of water and gradually raise the temperature. At approximately 10 deg C intervals, remove a pair of tubes, mix the contents and measure the rate of reaction and temperature as before. Plot a graph of the logarithm of the reciprocal of the time for the sulphur to appear against the reciprocal of the absolute temperature. This will be a straight line, showing that the relation $\log (\text{rate}) = A - B/T$ is followed (see Fig. 85).

Specimen result

Expt. 9b-2 The effect of temperature on the rate of decomposition
of sodium thiosulphate in acid solution

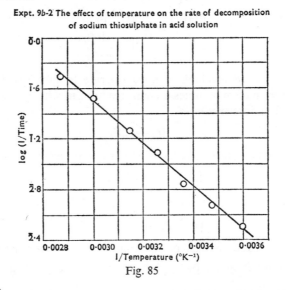

Fig. 85

Expt. 9b-3

Expts. 9a-3 and 4 (*q.v.*) may be repeated at temperatures 10 and 20 deg C higher than room temperature, and the temperature coefficients and energies of activation calculated. See specimen results under Expt. 9a-4.

(c) Catalysis

Expt. 9c-1

Note the effect on the rate of decomposition of potassium chlorate of addition of small quantities of (*a*) dried black copper oxide, (*b*) fine sand or powdered silica. Place about 1 g of the chlorate in each of two test-tubes and mix the catalyst into one; heat them side by side in the same moderate Bunsen flame and use a glowing splint to determine the effect of the catalyst on (*a*) the temperature at which oxygen is first evolved, and (*b*) its rate of production.

Expt. 9c-2

Mix a little powdered iodine with some aluminium powder on an asbestos card in the fume chamber. No reaction occurs. Add a drop of water. There is a vigorous reaction, heat being generated and causing the volatilization of iodine as clouds of violet vapour.

Expt. 9c-3

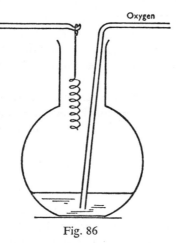

Oxygen

Fig. 86

Place a few cm^3 of 0·880 ammonia in a bolt-head or conical flask; warm it slightly and pass a slow stream of oxygen through it (see Fig. 86). Heat a spiral of platinum wire red-hot and hold it in the flask above the ammonia. The wire glows brightly as reaction takes place on its surface with evolution of heat, and striking, but harmless, explosions occur.

It has been convenient in the past to classify all substances that are not consumed by the chemical changes they facilitate as 'catalysts'. This generalization has effected a simplification which may be misleading, for the modes of action of these substances classed together as catalysts may differ widely from one case to another; the mechanism which may be shown to apply to one example may also explain others, but it does not follow that it applies to all.

There have been two major theories of catalysts: (1) the Intermediate Compound Theory, and (2) the Adsorption Theory, and the operation of most catalysts can be explained in terms of one or other of these theories.

(1) According to this mechanism, the reaction $A + B \longrightarrow D$, when occurring in the presence of the catalyst C, takes place in two stages, involving the formation of the intermediate compound AC: (i) $A + C \longrightarrow AC$, and (ii) $AC + B \longrightarrow D + C$. Each stage is faster than the direct reaction. For example, the reaction $2H_2 + O_2 \longrightarrow 2H_2O$, which is catalysed by freshly reduced copper, can take place via the formation of CuO: (i) $2Cu + O_2 \longrightarrow 2CuO$, and (ii) $2CuO + 2H_2 \longrightarrow 2H_2O + 2Cu$. The speed of a sequence of two reactions such as these is equal to the speed of the slower of the two. Since the net rate of the catalysed reaction is greater than that of the uncatalysed reaction, it follows that the slower of the reactions (i) and (ii) is faster than the direct reaction.

In a catalysed reaction taking place in accordance with this mechanism, the energy of activation (which is an important factor in determining the rate of the reaction) is that of the reaction $A + C \longrightarrow AC$ or of the reaction $AC + B \longrightarrow D + C$, whichever is the rate-determining step. They may well be of a lower value than the energy of activation of the direct reaction $A + B \longrightarrow D$.

(2) In many examples of heterogeneous catalysts, one or both of the reactants is adsorbed on the surface of the catalyst. That adsorption takes place can sometimes be shown directly, but its occurrence is borne out by the fact that the extent and condition of the surface of the catalyst greatly influence its activity. Larger surface area and greater breaking up of the surface into a more finely divided and irregular state increase the activity of the catalyst; adsorption of foreign

substances ('poisons') and a smoothing of the surface by sintering destroy the activity.

It is interesting to note that before the middle of the nineteenth century Faraday had stated that the seat of chemical change was the film of gas adsorbed at surfaces, and it is now known that many gas reactions take place in contact with a solid surface. Since the adsorbed molecules are held to the surface by large forces, they may be in a very different condition from free molecules in the gas phase. Various views of the nature of the adsorption complex have been put forward. In the extreme case surface compounds are formed between the adsorbed molecules and the surface atoms. Any distinction between the above two theories then becomes blurred. Gas molecules entering the strong electric fields existing at the solid surface are polarized and held to the surface by electrostatic forces. The degree of distortion thus produced in the molecule may be very great. The adsorbed molecules thus require a lower energy of activation for reaction than when they are in the gaseous phase. The rates of reaction at the surface are thus greater than those of the same reaction occurring in the gas phase. Many reactions once considered to take place only in the gas phase, in fact involve steps which occur on the surface of the containing vessel.

Expt. 9c-4

Fill a small gas jar (say c. 3–500 cm^3) with hydrogen from a cylinder or suitable generator, and close the jar with a lid.

Hold some freshly prepared platinized asbestos in a pair of tongs, warm it gently and hold it at the mouth of the gas jar immediately after removing the lid. The hydrogen-air mixture explodes. The platinized asbestos appears to be unchanged.

This example of catalysis is vividly described by Faraday (*Experimental Researches in Electrochemistry*, Everyman Edition).

A catalyst is a substance which affects the rate of a reaction, but is not itself changed chemically by the reaction. In the above experiments, manganese dioxide, copper oxide, water and platinum were acting as catalysts. Catalysts may be classified as: (i) homogeneous catalysts, such as hydrogen ions in, for example, the hydrolysis of esters or the inversion of cane sugar, etc., (ii) heterogeneous or surface catalysts, such as the vanadium pentoxide catalyst used in the contact process for sulphuric acid, the metal oxide catalysts used in the synthesis

EBC—O

of organic compounds from water gas, or the active nickel used in the hydrogenation of unsaturated hydrocarbons.

Expt. 9c-5   Homogeneous catalysis

The reaction between potassium dichromate and potassium iodide in acid solution to form iodine is catalysed by copper(II) ions. The course of the reaction may be followed in a manner similar to that used in Expt. 9a-4. The solutions required are: M/10-potassium dichromate, M/10-sodium thiosulphate, freshly prepared potassium iodide solution (about 5%), and starch solution.

To 10 cm³ of the dichromate solution in a conical flask add 2·5 cm³ of glacial acetic acid, 10 cm³ of water and a little starch. At zero time, 30 cm³ of the iodide solution is added and 1 cm³ of thiosulphate is run in from a burette. Note the time when the blue colour reappears and add another cm³ of thiosulphate. Continue in this way until about six readings have been obtained, and plot a graph of the number of cm³ of thiosulphate added (a measure of the extent of the reaction) against the time.

Specimen results

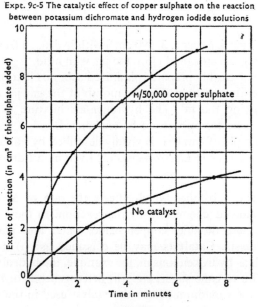

Expt. 9c-5 The catalytic effect of copper sulphate on the reaction between potassium dichromate and hydrogen iodide solutions

M/50,000 copper sulphate

No catalyst

Extent of reaction (in cm³ of thiosulphate added)

Time in minutes

Fig. 87

Repeat the experiment, but add 1 cm³ of M/1000-copper(II) sulphate solution before starting the reaction. It will be found that the rate is more than doubled. Repeat with the addition of 3 cm³ of the copper(II) sulphate solution, but start with 25 cm³ of dichromate and add the thiosulphate 3 cm³ at a time for the first few readings. The rate is greatly increased, hence the necessity for the larger quantities used.

Under the conditions described above, the reaction does not follow a simple first order law, so the effect of the catalyst on the rate is observed merely by inspecting the graphs. For another example of homogeneous catalysis see Expt. 9a-4 (iii), p. 185.

Expt. 9c-6   An autocatalytic reaction

Prepare about 10 cm³ of a saturated solution of sodium bisulphite (by shaking the solid with cold water—do not heat it). Pour the solution on to about 4 g of potassium chlorate in a litre flask. After a few minutes, a violent reaction takes place. This reaction is the reduction of chloric acid to hydrochloric acid. The weakly acidic sodium bisulphite acts on the chlorate to liberate some chloric acid. The latter oxidizes the bisulphite to the more strongly acidic bisulphate, and so the liberation of chloric acid from the chlorate is accelerated. The reaction between the original materials is thus autocatalytic, the catalyst produced being the bisulphate.

Expt. 9c-7   The reaction between potassium permanganate and oxalic acid

(An example of autocatalysis and of intermediate compound formation.)

The use of this reaction in volumetric analysis is well known. It is carried out at about 60 °C and goes rapidly to completion. However, if a dilute solution of potassium permanganate, in the presence of dilute sulphuric acid, is reduced by oxalic acid at room temperature, the reaction is slow and the colour changes from purple through sherry shades to pale yellow before the solution finally becomes colourless. The reaction is autocatalytic, the catalyst produced being manganese(II) ions. They function as a catalyst by causing the reaction to take place in two stages, with the intermediate formation of quadrivalent manganese.

The following reagents will be required: 1 litre each of 2M/500-potassium permanganate solution (M ≡ KMnO₄), 3M/500-manganese(II) sulphate solution, M/5-oxalic acid solution, M/200-sulphuric acid, and M/100-sodium thiosulphate solution. The solutions are best measured out from burettes. Small conical flasks are most suitable as reaction vessels. They should be washed out with a little permanganate, followed by distilled water, before use.

(i) *The reaction between potassium permanganate and oxalic acid solutions.* Put 5 cm³ of the permanganate solution and 5 cm³ of the M/200-sulphuric acid in each of ten flasks. At zero time, add 5 cm³ of the oxalic acid. After times t, varying from 0, 5, 10 to 60 min, add a little freshly made acidified potassium iodide solution. (The quantity added should contain about 10 crystals of potassium iodide the size of a rice grain and 2 or 3 cm³ of about M-sulphuric acid.) The reaction is thereby stopped, and its extent can be measured by estimating the iodine liberated by titration with M/100-thiosulphate, using starch solution as an indicator. Plot the 'titre' against the time of reaction t, and by measuring the slopes of the curves, plot the rate of reaction against the time.

(ii) *The same reaction in the presence of manganese(II) sulphate.* Repeat the above experiment, but together with the oxalic acid, add 5 cm³ of the manganese(II) sulphate solution. Plot graphs as before.

The results of Expt. (i) show that the rate of reaction increases at first to a maximum and then decreases. This is due to the formation of manganese(II) ions, which then act as a catalyst. Expt. (ii) illustrates this, in that the reaction begins at its maximum rate.

The following experiment elucidates the manner in which the manganese(II) sulphate acts as a catalyst.

(iii) *The reaction between acidified potassium permanganate and manganese(II) sulphate solutions.* Put 5 cm³ of the permanganate solution in each of six flasks, and add 5 cm³ of M/200-sulphuric acid to each. At zero time, add 5 cm³ of the manganese sulphate, and, after time t, ranging from $\frac{1}{2}$ to 3 min, stop the reaction by adding acidified potassium iodide solution and titrate the iodine liberated. Note that there is no change in the 'oxidizing power' of the solution as a result of the reaction.

Now repeat the above procedure, but after the permanganate and manganese sulphate have been reacting for time t, add 5 cm³ of the oxalic acid (previously measured out into test-tubes) and leave it to react for 10 sec in each case; then stop the reaction with potassium iodide and titrate the iodine liberated.

Expt. (iii) shows that the permanganate and the manganese salt react to give a brown precipitate (hydrated manganese dioxide). This substance reacts faster with the oxalic acid than does the permanganate. Hence the manganese sulphate functions as a catalyst via the formation of an intermediate compound containing quadrivalent manganese. The reactions may be represented by the equations:

$$2MnO_4^- + 5C_2O_4^= + 16H^+ \rightarrow 2Mn^{2+} + 10CO_2 + 8H_2O$$
$$A \quad + \quad B \quad\quad \rightarrow C \quad + \quad D$$

$$2MnO_4^- + 3Mn^{2+} + 16H^+ \rightarrow 5Mn^{4+} + 8H_2O$$
$$A \quad + \quad C \quad\quad \rightarrow AC$$

$$5Mn^{4+} + 5C_2O_2^= \rightarrow 10CO_2 + 5Mn^{2+}.$$
$$AC \quad + \quad B \rightarrow D \quad + \quad C$$

Expt. 9c-8 The rate of decomposition of hydrogen peroxide catalysed by manganese dioxide

The reaction is carried out in a conical flask fitted with a bung and tube which can be connected to a 100 cm³ glass gas-syringe. The volume of oxygen evolved is noted at half-minute intervals.

Place 2·5 to 3·0 cm³ of '20-volume' hydrogen peroxide, in the flask together with 25 cm³ of distilled water. Weigh 0·25 g of powdered manganese dioxide into a small test-tube and slip it, upright, into the flask. Insert the bung and connect the tube to the syringe. Tip up the flask so as to empty the manganese dioxide into the solution and at the same moment start a stop-clock. As the oxygen goes into the syringe, note the volume every half-minute. Allow the reaction to proceed to completion. The final volume of oxygen is a measure of a the initial concentration of the peroxide. After the reaction has been occurring for t seconds, the concentration of the peroxide is measured by $(a - x)$, where x is proportional to the volume of oxygen evolved at that time. Plot a graph of $\log\left(\dfrac{a}{a-x}\right)$ against t. See specimen result, Figs. 88 and 89.

Investigate the effect on the rate constant of changing (i) the initial concentration of the hydrogen peroxide; (ii) the quantity of manganese dioxide; (iii) the state of subdivision of the manganese dioxide.

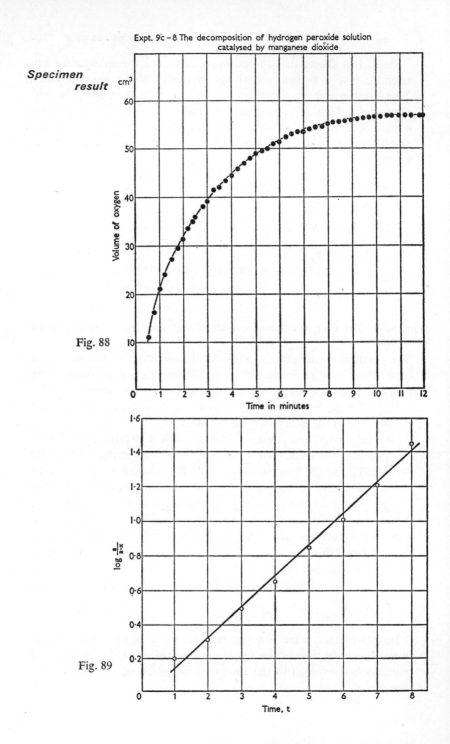

Expt. 9c – 8 The decomposition of hydrogen peroxide solution catalysed by manganese dioxide

Specimen result

Fig. 88

Fig. 89

10 Energy and chemical change

(a) Heats of reaction

Expt. 10a-1

Place 50 cm³ of water in a small Thermos flask or polythene bottle. Quickly weigh out approximately 3 g of phosphorus pentoxide on a filter paper, and drop the whole into the flask. Stir and note the rise in temperature on a thermometer reading to 0·1 deg C. Repeat with about 1 g of freshly heated quicklime, made by heating marble chips in a muffle furnace for at least an hour and allowing to cool to room temperature in a dessicator. Repeat again with about 3 g of anhydrous copper sulphate. The above weights are very roughly in the ratio of these of the moles of the three substances. Rises in temperature of about 14, 6 and 4 deg C respectively will be obtained.

Expt. 10a-2

In this experiment, the temperature changes that occur when (*a*) sodium hydroxide and (*b*) ammonium chloride, are dissolved in water are measured. Weigh out $\frac{1}{10}$ mole of sodium hydroxide pellets (4 g), and quickly tip it into 100 cm³ of water in a small Thermos flask or polythene bottle. Stir well and note the maximum temperature rise. Repeat the experiment using 0·1 mole of ammonium chloride (5·3 g). Note the maximum change in temperature.

Most chemical changes are accompanied by the evolution or absorption of heat. When heat is evolved, the reaction is termed 'exothermic' and when heat is absorbed, the reaction is termed 'endothermic'. For example, the combustion of carbon to carbon dioxide is exothermic, whereas the action of steam on carbon to form 'water gas' is endothermic. Heats of reaction are usually measured in kilocalories per mole, and refer to the number of moles of reactants indicated by the chemical equation for the reaction. Thus:

$$C + O_2 \rightarrow CO_2 \quad \Delta H = -97 \text{ kcal (at constant volume)}$$

i.e. when 12 g of carbon are burnt to carbon dioxide, 97 kcal are lost by the system to the surroundings; hence the negative sign.

$$C + 2S \rightarrow CS_2 \quad \Delta H = +25 \cdot 4 \text{ kcal}$$

i.e. when 12 g of carbon are converted to carbon disulphide, 25·4 kcals are absorbed by the system; hence the positive sign. If no other energy changes are involved in the reaction, the heat change is a measure of the change in internal energy (ΔU) of the system. But other energy changes may also occur, e.g. in a reaction in which a gas is evolved, work is done by the system as the gas pushes its way out into the surroundings. If this occurs at a constant pressure, p, and if the volume change caused by the reaction is ΔV, the work done against the external pressure is $p\Delta V$. Here the change in the internal energy of the system is equal to the sum of this quantity of energy and the quantity of heat energy evolved or absorbed. If the reaction is carried out at constant volume, no external work is done by the system and $\Delta U = \Delta H$.

If a reaction takes place in a voltaic cell, the energy released may be manifest partly as electrical energy. This subject is dealt with later (see pp. 222–236).

Changes in internal energy relate largely to the breaking of bonds and the making of other bonds between the atoms involved. Measurements of heats of reaction therefore give information about the energies stored in these bonds, i.e. about relative bond strengths. They are found to vary only slightly with temperature.

Among the heats of reactions that are most readily determined experimentally are: heats of combustion, heats of neutralization, and heats of precipitation.

Expt. 10a-3   Heat of neutralization

Determine the heat change when 1 mole of sodium hydroxide is completely neutralized by (*a*) hydrochloric acid (*b*) nitric acid. Carry out the reaction in a small Thermos flask or polythene bottle. Put 100 cm³ of a molar solution of sodium hydroxide in the flask or bottle and note the temperature. Measure 100 cm³ of molar hydrochloric acid in a measuring cylinder and adjust its temperature to be the same as that of the sodium hydroxide solution. Add the acid solution to the alkali, stir well and note the highest temperature reached. Take the specific heats of the solutions as unity, ignore the thermal capacity of the calorimeter, and calculate the heat evolved. Repeat the experiment using nitric acid. The heats of neutralization are about 13·7 kcal per mole in each case. How can you explain that they are about the same?

Expt. 10a-4 *Heats of combustion*

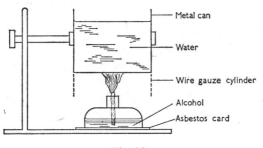

Fig. 90

As a simple calorimeter use a cylindrical metal tin can, approximately 6 cm in diameter and 8 cm high, to which has been soldered a cylinder of wire gauze, as shown in the diagram. Put 250 cm³ of water in the can.

Half fill a spirit lamp with ethanol and weigh the lamp with contents. Note the temperature of the water, light the lamp and place the calorimeter over it. Stir the water and when the temperature has risen exactly 30 deg C, remove the lamp and weigh it again. The loss in weight of alcohol burned and the heat given out per mole can be calculated. If the water equivalent of the calorimeter and heat losses to the atmosphere are ignored, the figure obtained for the heat of combustion will, of course, be too low.

Repeat the experiment using a series of alcohols, e.g. n-propanol, n-butanol, n-pentanol, n-hexanol, etc. When you have obtained values for the heats of combustion per mole, find the difference between each in the series. Are they similar? How can this be explained in terms of the structure of the molecules?

Expt. 10a-5 *Heat of the reaction Zn + $Cu^{2+} \rightarrow Zn^{2+} + Cu$*

Use a polythene bottle or small Thermos flask as the calorimeter. Place it in exactly 50 cm³ of an M/5 solution of copper(II) sulphate and read the temperature to 0·1 deg C. Add a little more than 1 g of zinc dust, mix thoroughly and note the maximum rise in temperature. Assuming the specific heat of the solution to be unity, neglect the heat capacity of the calorimeter and the metals, and calculate the heat evolved in the reaction. Express your result in kcal per mole of reaction. Compare with the result of Expts. 11b–3,4.

Expt. 10a-6   Heat of precipitation

It is interesting to compare the heats given out when one mole of each member of a series of similar compounds is precipitated. The following are suitable (a) silver halides, (b) ferrocyanides of heavy metals, e.g. zinc, copper, nickel, cobalt, iron, (c) carbonates of magnesium, calcium, strontium and barium. Start with 0·01 mole of one of the reactants and mix with rather more of the other reactant in each case. Use the method of the previous experiment and compare the heats of reaction per mole in each series.

Specimen result

Heats of precipitation of alkali metal carbonates

Magnesium $\Delta H = + 3·4$ kcal per mole
Calcium $\Delta H = + 2·2$,,   ,,   ,,
Strontium $\Delta H = + 0·7$,,   ,,   ,,
Barium $\Delta H = - 1·2$,,   ,,   ,,

HEATS OF FORMATION

The heat of formation of a compound is defined as the number of calories evolved or absorbed at constant volume when 1 mole of the compound is formed from its elements in their normal state, i.e. as they exist at room temperature. If the reaction leading to the formation of a compound from its elements in their normal state is exothermic, the compound is called an exothermic compound. If heat is absorbed in the formation of the compound, it is called an endothermic compound. For example, carbon dioxide is an exothermic compound whereas carbon disulphide is endothermic.

Heats of formation are an indication of the relative energy contents of substances. Thus it is clear that 1 mole of carbon disulphide contains 25·4 kcal more than 1 mole of carbon plus 2 moles of sulphur. Again, the energy content of 1 mole of carbon dioxide is 97 kcal less than the sum of the energy contents of 1 mole of carbon and 1 mole of oxygen.

The heat of formation is also some indication of the stability of the compound. As we might expect, endothermic compounds are often very unstable and most stable compounds are exothermic. It is seldom possible to measure the heat of formation directly, for only a limited number of compounds can be formed directly from their elements. However, a series of reactions can usually be found which, starting with the elements in their normal forms, leads through two or more steps to the formation of the compound. If the heat changes at constant volume in all these reactions can be measured, the heat of formation can be calculated. Such a calculation assumes the law of conservation of energy, which, in this particular application, is expressed by Hess's 'Law of Constant Heat Summation'. This law states that 'the heat evolved in any chemical change is independent of the manner in which it is carried out, whether in one or more steps'. As an example, consider the determination of the heat of formation of carbon monoxide. Carbon cannot be burnt directly to carbon monoxide only, so Hess's law is used. The heat of combustion of carbon to carbon dioxide is given by

$$C + O_2 \rightarrow CO_2 \quad \Delta H = -97 \text{ kcal.}$$

The heat of combustion of carbon monoxide is given by

$$CO + \tfrac{1}{2}O_2 \rightarrow CO_2 \quad \Delta H = -68 \text{ kcal.}$$

Let the heat of formation of carbon monoxide be Q. Then

$$C + \tfrac{1}{2}O_2 \rightarrow CO \quad \Delta H = Q.$$

By Hess's law, $\qquad Q - 68 = -97$

$$\therefore Q = -29 \text{ kcal.}$$

In this way, Hess's law can be used to derive the heats of reaction of many reactions which could not be measured directly.

Heats of formation (and other heats of reaction) may vary with the conditions of temperature and pressure under which they were measured. The symbol $\Delta H°$ is therefore used to specify the value of ΔH under 'standard' conditions. These are room temperature (often specified as, say, 25°C), one mole of a gas at one atmosphere, and unit activity for solutions (i.e. approximately 1 mole per litre in dilute solution).

Expt. 10a-7 *Illustrations of Hess's law*

(1) The reaction between solid sodium hydroxide and dilute hydrochloric acid,

$$\text{NaOH (solid)} + \text{HCl (solution)} \rightarrow \text{NaCl} + H_2O \qquad \Delta H = q_1$$

could be carried out in the following two steps:

$$NaOH \text{ (solid)} + water \rightarrow NaOH \text{ (solution)} \qquad \Delta H = q_2$$

$$NaOH \text{ (solution)} + HCl \text{ (solution)} \rightarrow NaCl + H_2O \quad \Delta H = q_3$$

According to Hess's law, $q_1 = q_2 + q_3$.

A value for q_2 was obtained in Expt. 10a-2, for q_3 in Expt. 10a-3, so it remains to measure q_1.

Place 100 cm³ of M-hydrochloric acid in a small Thermos flask or polythene bottle and note the temperature of the acid to 0·1 deg C. Weigh out 4 g of solid sodium hydroxide pellets and add this to the acid. Stir well and note the maximum rise in temperature. Calculate the heat evolved in kcal per mole of reactants. To what extent do your results accord with Hess's law?

(2) Another reaction which can easily be carried out in two stages is that between lithium (or calcium) and dilute acid.

$$Li + HCl \rightarrow LiCl + \tfrac{1}{2}H_2 \qquad \Delta H = q_1$$

The two stages are $\quad Li + H_2O \rightarrow LiOH + \tfrac{1}{2}H_2 \quad \Delta H = q_2$

and $\qquad\qquad\qquad LiOH + HCl \rightarrow LiCl + H_2O \quad \Delta H = q_3$

Put 50 cm³ of water in a small Thermos flask or polythene bottle and note the temperature. Cut a cube of lithium about 3–4 mm wide, wrap it in a small piece of lead foil in which holes have been pierced, and drop it into the water. Stir well and note the maximum temperature reached. Keep the solution for the next part of the experiment.

To find q_3, put 50 cm³ of M-hydrochloric acid in a measuring cylinder and note its temperature. Pour the lithium hydroxide solution obtained in the first part of the experiment into a small flask and adjust its temperature to the same value as that of the acid. This is best done by immersing the flask in ice and water until the temperature is a little too low, and then warming the flask with the hands. Then pour both solutions into the vessel being used as a calorimeter and note the final temperature.

The quantity of acid added will have been more than enough to neutralize all the lithium hydroxide. Titrate the excess hydrochloric acid with M-sodium hydroxide solution, and hence calculate the number of moles of lithium hydroxide that were present. This is probably a more accurate way of determining the quantity of lithium used than by weighing the lithium metal. The value of q_2 can now be calculated.

To find q_1, use a similar quantity of lithium and react it with 50 cm³ of M-hydrochloric acid using the procedure described above for the reaction of the metal with water. Again determine the quantity of acid remaining by titration with molar sodium hydroxide solution. Hence calculate the number of moles of lithium used and the heat of reaction q_1, in kcals per mole. To what extent are your results in agreement with Hess's law?

(b) Chemical affinity

Expt. 10b-1

(1) Hold a piece of magnesium ribbon with a pair of tongs and heat in a Bunsen flame. Repeat with a piece of copper foil.

(2) Place a mixture consisting of about as much magnesium powder as would cover a sixpence and an equal quantity of dry copper oxide in a crucible; put a Bunsen flame beneath it and stand back.

It is commonplace that elements and compounds vary greatly in their chemical reactivity. For example, magnesium oxidizes very rapidly when heated in the air, but copper only slowly. Copper oxide is easily reduced by hydrogen, magnesium oxide is not reduced. Magnesium metal, when mixed with copper oxide and heated, reduces the latter and magnesium oxide is formed. Under these conditions it is clear that magnesium has a greater affinity for oxygen than copper has.

The fact that chlorine displaces bromine, and that bromine displaces iodine from solutions may be quoted as another example of relative chemical affinity. Hydrogen chloride is more stable to heat than hydrogen bromide, which in turn is more stable than hydrogen iodide, indicating the decreasing affinity of chlorine, bromine and iodine for hydrogen. In the eighteenth century, attempts were made to compare affinities in this sort of way and tables of affinity were drawn up.

On this view of the matter, if the reaction $A + BC \rightarrow AB + C$ takes place because the affinity of A for B is greater than that of C for B, the reverse reaction should not occur at all. Thus when Berthollet, in 1801, established the reversible nature of many reactions, this simple theory of chemical affinity was shown to be inadequate.

An important contribution to the theory of chemical affinity was made by Julius Thomsen in 1854. He regarded the heat liberated in a reaction as a measure of its affinity on the grounds that greater evolution of heat accompanies reactions that proceed more readily and that lead to the formation of more stable compounds. Although, as we shall see, this view cannot be correct, heats of reaction are a useful guide for the comparison of the affinities of reactions, and heats of formation are a useful indication of the relative stabilities of compounds. However, on this view, endothermic reactions, i.e. those in which heat is absorbed, should have a 'negative affinity' and should not occur.

In order to answer the question 'Why do chemical reactions occur?', further consideration must be given to the energy changes that accompany reactions. It has already been stated (p. 200) that the energy released in a chemical change as a result of the net change in bond energies can appear as heat or as other forms of energy. The difference in the energy contents of reactants and products (provided their temperatures and pressures are the same) is independent of the form in which the energy appears. Let ΔU represent the change in internal energy as a result of the breaking of bonds and the making of new ones. In a reaction in which a gas is evolved at a constant pressure p, involving a volume change ΔV, the energy lost by the reacting substances to the surroundings is equal to the energy gained by the latter. This energy comprises thermal energy plus the energy resulting from the work done by the system on the surroundings, i.e. $p\Delta V$.

So
$$-\Delta U = \Delta H + p\Delta V$$
or
$$-\Delta H = \Delta U + p\Delta V$$

i.e. the heat energy that appears is equal to the sum of the changes in internal energy and the external work done. This quantity, $(\Delta U + p\Delta V)$, represented by the symbol ΔH, is known as the *enthalpy* of the reaction. It is the heat evolved or absorbed in the reaction at constant pressure and is often called the heat of reaction. Most of the heats of reaction measured in the preceding section are enthalpies.

If the reaction takes place at constant volume, $p\Delta V = 0$ and $\Delta H = \Delta U$. In this case, the heat of reaction measures the changes in bond energies. But if $p\Delta V$ is not zero, the heat or enthalpy of reaction exceeds the change in internal energies by an amount $p\Delta V$. Since most reactions are carried out at atmospheric pressure, heats of reaction, heats of formation, etc. are perhaps better referred to as enthalpies. However, $p\Delta V$ is often quite small and the enthalpies of reactions involving volume changes are a close approximation to the heats of reaction at constant volume.

Reactions may take place, not only as discussed above, in a calorimeter, but in a voltaic cell. For example, if the reaction between zinc and copper(II) sulphate solution is carried out in a calorimeter (as in Expt. 10a-5) the energy released appears as heat, but if the reaction is carried out in a voltaic cell part of the energy can be caused to give rise to an electromotive force. If a current is being drawn from the cell, the reaction is taking place and some energy is released as heat and some as electricity. The larger the current, the greater the proportion of energy released as heat. To get the maximum electrical work from the reaction, it must take place extremely slowly so as to give only an infinitesimal current. In the limit, the reaction is not occurring at all. This limiting amount of work a reaction could yield is known as the *free energy* of the reaction and is symbolized by ΔG. If the conditions are 'standard' (see p. 203), the value of ΔG is symbolized by $\Delta G°$. In most reactions $\Delta G°$ is not the same as the enthalpy change, $\Delta H°$, but the energy appearing as heat and that appearing as work must always add up to $\Delta H°$.

The following diagram (after H. F. Halliwell) illustrates the relationship.

Suppose that ΔH for a reaction is -20 kcals per mole, there are many possibilities:

	(1) All the energy changes appear as heat	(2) Some energy appears as work	(3) Maximum amount of work done either or	
Initial state				
ΔH	20 kcals as heat (Exothermic)	8 kcals as heat (Exothermic) 12 kcals as work	2 kcals as heat (Exothermic) 18 kcals as work (maximum)	(Endothermic) 25 kcals as work (maximum)
Final state				5 kcals put back as heat

The difference between ΔH and ΔG is defined as the product of the absolute temperature, T, and the change in a quantity called the *entropy*, S. Hence, $\Delta H = \Delta G + T\Delta S$.

One way of regarding entropy is as a measure of the degree of randomness of the system. To get some idea of the meaning of entropy, carry out the following simple experiment.

Expt. 10b-2

Place 50 cm³ of cold water in a small Thermos flask or plastic bottle, and 50 cm of nearly boiling water in another flask or bottle. Put a thermocouple in each sample of water and connect the thermocouple in series with a galvanometer. It is clear that energy can be obtained from this system of hot and cold water in the form of an electric current. Now pour all the water into one vessel, mix well, and return half to each Thermos flask. Again insert the thermocouples. The possibility of obtaining electrical energy has now disappeared.

When the hot and cold samples of water were mixed, no heat was lost from the system. i.e. $\Delta H = 0$, but there was a decrease in the free energy from a value G to zero. At the same time the degree of randomness increased because the particles of water, belonging to either a hot, or a cold group at the start, after mixing, all belonged to the same lukewarm group. $T\Delta S = -\Delta G$.

The sign of $\Delta G°$ indicates whether a reaction will take place or not. For example, the reaction $Zn + Cu^{2+} \rightarrow Zn^{2+} + Cu$ can supply energy when occurring in a voltaic cell (the Daniell cell). The system loses energy so $\Delta G°$ is negative.

$\Delta G°$ may be obtained from measurement of the e.m.f. of a cell in which the reaction could occur, so such measurements give information about the feasibility of a given reaction, i.e. whether it will occur or not. If $\Delta G°$ is negative, the reaction could go.

In the case of an exothermic reaction, in which $\Delta H°$ is also negative, this means that, provided there is not a large decrease in entropy, the reaction could go:

$$\Delta G° = \Delta H° - T\Delta S°$$

In the case of an endothermic reaction, in which $\Delta H°$ is positive, it could take place if there is a sufficiently large increase in entropy to make $\Delta G°$ negative. In this way, the occurrence of endothermic reactions is explained—they occur because the relatively large increase in entropy overcomes the absorption of heat energy, thus still enabling the reaction to do work (i.e. making $\Delta G°$ negative).

As an example, consider the formation of water gas:

$$C + H_2O \rightarrow CO + H_2 \quad \Delta H° = + 31 \cdot 4 \text{ kcal.}$$

For this reaction to go, therefore, $T\Delta S°$ must be at least $31 \cdot 4$ kcal. It is known that the entropy change for 1 mole of these reactants is 32 calories per degree.

At 300°K, $T\Delta S° = 300 \times 32 = 0 \cdot 96$ kcal, i.e. less than $31 \cdot 4$ and the reaction will not occur. However, at 1000°K, $T\Delta S° = 1000 \times 32 = 32$ kcal, i.e. greater than $31 \cdot 4$ and so the reaction will go at a temperature of 1000°K or above.

For another example, see p. 277, 'Solubility'.

11 Electrochemistry

(a) Electrolysis and the theory of ions

Expt. 11a-1

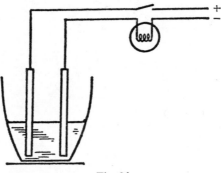

Fig. 91

Connect a lamp-holder and lamp in series with two copper-wire electrodes, a switch and appropriate battery (see Fig. 91). Melt some potassium or cadmium iodide, or lead bromide, in a crucible or pyrex beaker, insert the electrodes and allow the salt to cool and solidify. Switch on and note that no current passes. Melt the salt again and note that the lamp lights, showing that the salt has become conducting. Iodine or bromine vapour is evolved at one of the electrodes, indicating that the salt is being decomposed.

Repeat the experiment using sugar or sulphur instead of salt, and note that no conduction or decomposition takes place.

The above experiment shows that liquids, like solids, can be classified as conductors or non-conductors of electricity. The conduction of electricity by liquids obeys Ohm's law, but differs from conduction by solids in that the passage of the current through the liquid is invariably accompanied by chemical decomposition. To this process Faraday gave the name 'electrolysis'. In his *Experimental Researches in Electricity*, he says: 'All the substances that conduct are such as could be decomposed by the electric current.'

Faraday found that many substances, although non-conductors when solid, became conducting when melted. He asked: 'Does solidification prevent conduction merely by chaining the particles in their place and preventing their separation in the manner necessary for decomposition?' Those substances that became conductors on melting he called 'electrolytes', and those that did not, 'non-electrolytes'.

Faraday introduced a number of electrochemical terms, most of which have been used ever since. The terms 'anode', 'cathode', 'anion' and 'cation' are due to him. Faraday realized both the useful and the possibly misleading aspect of new technical terms, for he wrote that he was 'fully aware that names are one thing and science another'.

Faraday interpreted the phenomena of electrolysis in terms of the ionic theory. He introduced the term 'ion' for 'those bodies that pass to the electrodes during electrolysis'. Some substances are made up of aggregates of these electrically charged particles, and it is these that are responsible for carrying the current during electrolysis. When such a substance is melted, or dispersed in solution, the ions are released from the crystal and move about with random thermal motion. If an electrical potential difference is then applied across a section of the liquid, the ions will migrate, the positive ions in the direction of lower positive potential and the negative ions in the opposite direction. Faraday proposed the term 'electrode' for 'the substance, or rather surface, whether of air, water, metal or any other body, which bounds the extent of the decomposing matter in the direction of the electric current'. He termed the positive pole the 'anode' and the negative pole the 'cathode'. The positive ions he called 'cations', since they move towards the cathode, and the negative ions are called 'anions'.

The ions are discharged at the electrodes, and it is here that the chemical decomposition occurs. This may simply be the liberation of the discharged ions, e.g. the deposition of copper at the cathode in the electrolysis of a copper salt, or it may also involve reaction between the ions and the electrode or the electrolyte. Thus in the electrolysis of sodium chloride in the Castner-Kellner cell, the sodium reacts with the mercury electrode forming an amalgam, and the chlorine reacts to some extent with the electrolyte forming a hypochlorite.

In addition to a qualitative study of many examples of electrolysis, Faraday made a quantitative study of the amount of chemical decomposition accompanying the passage of the current. He found that the amount of hydrogen liberated in the electrolysis of acidulated water was 'a very good measurer' of the quantity of electricity passed. He used this to investigate the effect of current strength, the nature, shape and size of the electrodes, etc. He summed up his results in what he called the 'law of definite action', now known as Faraday's Laws of Electrolysis. His wording of the laws was:

(1) 'The chemical power of a current of electricity is in direct proportion to the absolute quantity of electricity which passes.'
(2) 'For a constant quantity of electricity, the amount of electrochemical action is also a constant quantity, whatever the decomposing conductor may be.'

Faraday drew attention to the quantitative results of his measurements: 'Now it is wonderful to observe how small a quantity of a compound body is decomposed by a certain portion of electricity.' As an example he says that 'one grain of water, for decomposition, requires a current powerful enough to maintain a platinum wire, $\frac{1}{104}$ in. thick at red heat for $3\frac{3}{4}$ min'. Again, if charges equal to that associated with 1 mg of hydrogen were given to two small spheres placed 1 m apart, they would repel each other with a force of a million million tons. The charges concerned in the most brilliant flash of lightning will not decompose a single drop of water.

THE LAWS OF ELECTROLYSIS

When a given electrolyte is decomposed by the passage of an electric current and the quantities of matter decomposed by various quantities of electricity are measured, it is found that

$$m = kct$$

where m is the mass of substance decomposed by a current c flowing for time t, and k is a constant characteristic of the electrolyte.

For example, the mass of copper, m_{Cu}, deposited on the cathode in the electrolysis of copper sulphate solution is given by the equation $m_{Cu} = k_{Cu}ct$, where k_{Cu} is the mass of copper deposited by unit current flowing for unit time, and is known as the 'electrochemical equivalent of copper'. This relation expresses Faraday's first law: 'The mass of substance decomposed by electrolysis is directly proportional to the quantity of electricity passed.'

When the values of the electrochemical equivalent of, say, copper and silver, k_{Cu} and k_{Ag} are measured and compared, it is found that they are in the ratio of the chemical equivalents of the elements, viz. $k_{Cu} : k_{Ag} :: E.W._{Cu} : E.W._{Ag}$. This relationship is expressed in Faraday's second law, which in its modern form states that 'the masses o different chemical substances decomposed by equal quantities of electricity are in the ratio of their chemical equivalent weights'.

The quantitative results may be summed up in the statement that 1 gram-equivalent of any electrolyte is decomposed by 96,540 coulombs of electricity. The quantity of electricity which will deposit one mole of a univalent metal will deposit half a mole of a divalent metal. This quantity of electricity, 96,540 coulombs, is called one Faraday.

Faraday's laws of electrolysis are to the ionic theory as the laws of definite, multiple and equivalent proportions are to the atomic theory. A very complete analogy was drawn by Helmholtz in 1881 between chemical compounds and Faraday's ions regarded as compounds between matter and electricity. Just as chemical compounds differ in their properties from their constituents, so ions differ in properties from neutral atoms; for example, sodium is a soft, silvery metal, reacting violently with water, whereas 'sodium-ion' is colourless and soluble in water without decomposition.

The second law of electrolysis shows that, whilst all ions are built up in accordance with the law of constant composition, not every ion carries the same charge. The silver ion has just half the charge of a zinc ion. Why then does zinc displace silver from silver nitrate solution? According to Faraday, not because zinc combines with more electricity than silver, but because zinc ion is a more stable compound than silver ion. The idea of constant composition as applied to ions gives a useful interpretation of the discharge of an ion. When a positively charged fragment of expanded polystyrene is attracted by a rubbed polythene or bakelite rod, the positive charge on the polystyrene is first neutralized by the negative charge on the rod. This is followed by the negative charging of the polystyrene and its repulsion. Why does this not happen to a silver ion on reaching the cathode? Because the ion is a compound and not a mixture.

Faraday himself showed that electrical charge, like matter, is incapable of being created or destroyed (the ice-pail experiment). Electrical charge also follows the laws of multiple and reciprocal proportions:

the ratio of the weights of iron(III) to iron(II) which will separately combine with 96,540 coulombs of positive electricity is exactly 2 : 3; 32 g of zinc and 17 g of hydroxyl will separately combine with 96,540 coulombs of electricity, and 32 g of zinc combine with 17 g of hydroxyl. Here, then, we have electricity obeying four of the classical chemical laws of matter. When we are concerned with matter, we take these as the basis of the atomic theory. There is no logical reason for not taking the same laws as the basis for an atomic theory of electricity. It is remarkable that so early as 1834, Faraday almost reached the same position, for he writes: 'The harmony which this theory of the definite action of electricity introduces into the associated theories of definite proportions and electrochemical affinity, is very great . . . if we adopt the atomic phraseology, then the atoms of bodies which are equivalents to each other in their ordinary chemical action, have equal quantities of electricity naturally associated with them' (1834). Considering that the atomic theory was not logically established until 1858, Faraday's caution was more than justified. So far as he knew, there was precisely as much evidence for the electron as there was for the atom.

By 1881, however, Helmholtz could afford to be bolder. 'We cannot avoid concluding that electricity also, positive as well as negative, is divided into definite elementary portions, which behave like atoms of electricity.' Lord Kelvin referred to this statement as 'an epoch-making monument of the progress of natural philosophy'. Indeed, 1881 is the date of the discovery of the electron.

Expt. 11a-2   Faraday's laws of electrolysis

(1) Cut two electrodes from a piece of stout, pure copper foil, about 2 by 4 cm, and connect them in series with a milliammeter and variable resistance to a 4 V supply. Support them in a beaker containing an approximately 10% solution of pure copper(II) sulphate in distilled water. Clean the foil that is to be the cathode by rubbing it with fine emery paper and cotton wool under running water; finally wash it with alcohol, dry it some distance above a Bunsen flame, and weigh it to three decimal places. Pass a current of 20 mA for 5 hr, remove the cathode and, holding it by the upper corner with forceps, wash it by gently pouring distilled water, and then alcohol, over it. Dry it as before and weigh it again.

Repeat the experiment using a current of different strength, say 10 or 30 mA, passing for the same time, and compare the weights of copper deposited.

(2) Join a second cell in series with the above, but with 10% silver nitrate as the electrolyte and with silver foil electrodes. Compare the masses of copper and silver deposited by the same quantity of electricity. This experiment may, of course, be performed concurrently with (1).

Expt. 11a-3 The law of multiple proportions for electricity

Set up two voltameters as in the previous experiment and arrange for one of the beakers to be heated. Use copper foil electrodes in both but this time weigh the *anodes* before and after the experiment. Use 0·5M-copper(II) sulphate solution in the first voltameter and a solution containing 100 g of sodium chloride and 1 g of sodium hydroxide per litre in the other. Maintain the temperature of the latter at 80°C. Use a 6-volt battery and pass a current of 10–20 mA per cm³ of anode area for 15 to 20 minutes. A yellow-brown substance will form at the anode of the cell containing the alkaline electrolyte. This is copper(I) oxide. Compare the losses in weight of the two anodes. Copper is dissolving from one in the form of Cu^{2+} ions and from the other as Cu^+ ions and the losses in weight from the two anodes can be shown to be in the ratio 1 : 2 within a few per cent.

THE RELATIVE STABILITY OF IONS

Expt. 11a-4

Make fairly concentrated solutions of the following salts in distilled water in boiling-tubes:

Magnesium sulphate   Lead nitrate
Zinc sulphate   Mercury(II) chloride
Iron(II) sulphate   Silver nitrate
Copper sulphate

Obtain small specimens of as many of the following metals as is practicable, preferably in the form of foil or strip:

Magnesium   Iron   Mercury
Zinc   Copper   Silver
      Lead

Place the sample of iron in the copper sulphate solution and note that copper is displaced from solution by the iron. In a similar manner observe which of the others each metal in turn will displace from solution.

When metallic iron displaces copper from copper(II) sulphate solution forming a red deposit of copper and leaving a green solution of iron(II) sulphate, the reaction occurring is represented by the change:

$$Cu^{2+} + Fe \rightarrow Cu + Fe^{2+}.$$

So the behaviour of the metals in the above experiment is related to the tendency of each metal to lose electrons and form positive ions:

$$M \rightarrow M^{2+} + 2e.$$

Thus we may deduce that iron has less affinity for electrons than copper and more readily forms positive ions. The metals can be arranged in a series such that any metal will have a greater tendency to ionize than those that follow it (see table, p. 227). The elements at the top of the series are said to be more *electropositive* than those at the bottom. This does not mean that a charged atom of iron has more positive charge than a copper ion (indeed, they have the same charge) but that the iron atom does not so readily part with that charge. The phrase 'more electropositive' is therefore a reference to the degree of stability of the ion, not to its electrical charge.

Expt. 11a-5

Using strips of the metals listed in Expt. 11a-4, arrange a small cell containing dilute sulphuric acid so that two metal strips can be supported at a fixed distance apart. Connect the strips by copper wire to an insensitive galvanometer. (A suitable instrument can be made by winding a dozen turns of covered copper wire (about s.w.g. 32), fixing it vertically in the meridian, and supporting a pocket compass on a plasticine pillar in the centre.) Start with the copper and iron strips and determine the direction and approximate magnitude of the deflexion of the galvanometer needle. Work through the metals in pairs, and arrange them in an order such that any metal is more electropositive than those following it. (N.B. The terms used may be confusing if one does not remember that when two metals are arranged in a cell as above, the more electropositive metal, having a greater tendency to form positive ions, will acquire a *negative* potential with respect to the other metal.) With a little

ingenuity it is possible to place mercury, magnesium, and even
sodium in the series by means of this method.

In order to explain these electrochemical phenomena, it is assumed
that when a metal is immersed in water, it tends to throw off positive
ions into the water. This tendency was called by Nernst the *electrolytic
solution pressure* of the metal. It results in the formation of a layer of
ions around the metal, for the positive ions from the metal do not
migrate to distant parts of the solution, but are retained near the metal
by the negative charge it has acquired as a result of parting with
positive ions. The formation of this layer prevents further ionization
of the metal, and, as we shall see later, also largely prevents the metal
from attracting 'foreign' positive ions from the solution. If the metal
is immersed in a solution already containing some of its own positive

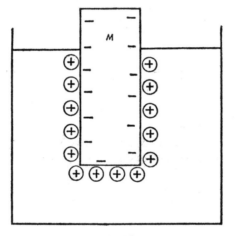

Fig. 92

ions, its tendency to ionize will be opposed by the tendency of these
ions to deposit on it. Nernst assumed that an equilibrium is set up
between these two tendencies. The former is measured by the electrolytic solution pressure of the metal, P, and the latter by the bombardment pressure of the solute ions, p. (For weak solutions this is approximately equal to the osmotic pressure of the solution, but these two
quantities are not the same.) A potential difference is thus set up between
the metal and the solution, and is called the *electrode potential*.

If P is greater than p, the metal will acquire a negative potential with respect to the solution of its ions, but if P is less than p, the metal will be positive with respect to the solution.

Electrode potentials thus vary from one metal to another owing to the differing values of P, and also vary with the concentration of the electrolyte because the latter affects the value of p. See table on p. 234.

GALVANIC COUPLES

Expt. 11a-6   The decomposition of water by a galvanic couple

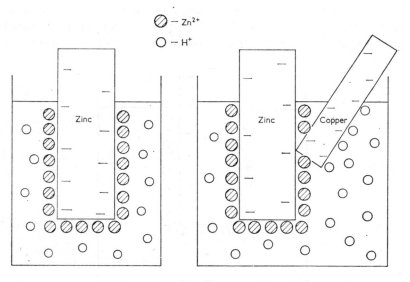

Fig. 93

Put a piece of pure zinc in dilute sulphuric acid. Touch the zinc with a piece of platinum or copper wire. Notice how the speed of evolution of hydrogen is affected, and where the gas is evolved (see Fig. 93).

The zinc throws off positive zinc ions into the solution and the metal acquires a negative potential with respect to the solution. The zinc ions are not regarded as free to move about away from the zinc, but as held as a layer on the zinc by the negative potential field of the metal. This layer repels positive hydrogen ions, which are thus only slowly, if at all, neutralized by the zinc and liberated as hydrogen. When the zinc is touched by platinum or any metal less electropositive than hydrogen, the negative charge spreads to the platinum and there neutralizes hydrogen ions, which come off as hydrogen gas. Impure zinc reacts more rapidly than the pure metal with dilute acid owing to the existence of patches that are less electropositive than the bulk of the metal. These patches act in a similar manner to the platinum in the above experiment and form a number of 'galvanic couples' with the zinc. A galvanic couple may be regarded as a voltaic cell in which the connection between the electrodes by an external circuit is replaced by direct contact of the metals within the electrolyte.

Among metallic couples which can be used to decompose water may be mentioned copper and zinc, and aluminium and mercury. The former is prepared by immersing granulated zinc in copper sulphate solution and washing with water. It will liberate hydrogen from boiling water without any acid. The aluminium amalgam will decompose cold water.

CORROSION

When pairs of unlike metals joined together are exposed to moisture, corrosion is liable to occur. The presence of an electrolyte such as carbonic acid or brine accelerates the process and the term 'electrolytic corrosion' is used to describe the phenomenon. When electrolytic corrosion occurs, the more electropositive metal suffers, the more electronegative metal being protected thereby. For example, when galvanized iron begins to corrode, the zinc coating is sacrificed and the area of corrosion spreads considerably before the iron itself is seriously affected. On the other hand, when tin-plate corrodes, the attack penetrates into the iron, forming corrosion pits and eating holes through the plate. The use of zinc protectors on the steel hulls for ships is intended to save the steel from electrolytic corrosion due to the bronze propellers (see Fig. 94).

The corrosion product may protect the metal from further attack, for example, the oxide film formed on aluminium is compact and tenacious and is responsible for the high resistance to corrosion shown by aluminium. It is possible to increase this oxide film by treating the aluminium with an oxidizing agent (e.g. an acid chromate solution) or by anodizing, and thus to improve the corrosion resistance of the metal.

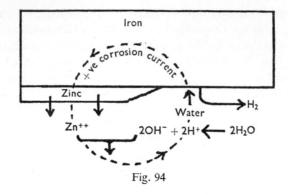

Fig. 94

The rusting of iron

The rusting of iron is also an electrolytic process. The conditions necessary for rusting to occur are as follows: iron only rusts in the presence of water; furthermore, the water must be condensed on the surface of the metal as liquid and must contain some electrolyte. Rusting can be prevented by warming the metal above the dew-point or by coating it with paint or with another metal, e.g. zinc or tin.

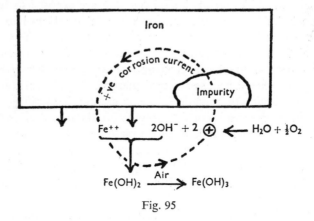

Fig. 95

Impurities in the iron form little galvanic couples with it, and if the impurity is less electropositive than the iron, the reactions illustrated in Fig. 95 will occur. Hydroxyl ions are produced in the solution by dissolved oxygen at the less electropositive patches in the metal. These combine with the iron(II) ions thrown off from the iron to form colloidal iron(II) hydroxide, which is soon oxidized to iron(III) hydroxide.

Thus rust itself is a secondary product and, being formed in the solution, affords no protection to the iron surface against further attack. The electrical circuit is completed through the liquid film on the iron which must therefore contain a dissolved electrolyte if rusting is to occur. This may be provided by carbon dioxide dissolved from the air, or by other electrolytes, e.g. acid fumes or salt spray from the sea. The process is illustrated by the following experiments.

Expt. 11a-7

(1) Clean a sheet of iron with emery paper. Place in the centre a large drop of a solution of sodium chloride containing a little potassium ferricyanide and phenolphthalein. The edge of the drop soon goes pink, showing the formation of hydroxyl ions where the oxygen has free access, and a blue precipitate forms in the liquid showing the presence of iron(II) ions. After a time, a ring of rust forms inside the edge of the drop.

(2) Suspend a clean strip of iron in a dilute salt solution, and notice that the rust forms beneath the surface of the liquid and not above it, i.e. rust forms where the oxygen concentration is least.

Expt. 11a-8

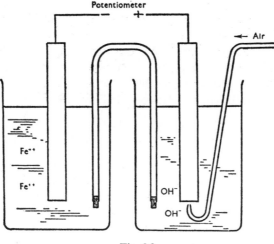

Fig. 96

Set up two half-cells with iron electrodes and dilute sodium chloride as electrolyte, connect them to a potentiometer or high resistance voltmeter and blow air over one electrode. Note that this electrode becomes electropositive with respect to the other. Leave the cell short-circuited for some time and test for iron(II) ions in the half through which air was not blown and for hydroxyl ions in the other (see Fig. 96).

The last experiment shows that rusting is promoted by a non-uniform distribution of oxygen over the iron. Those areas where there is most oxygen acquire a positive potential with respect to the others, behaving like patches of a noble metal, and the rust forms where the oxygen concentration is least.

(b) Voltaic cells

Expt. 11b-1

Make a stack of discs of copper foil, filter paper wet with dilute sulphuric acid, and zinc sheet, repeated *ad lib*. Connect the two end discs, one of copper and the other of zinc, to a voltmeter.

Expt. 11b-2

Set up a plate of copper and a plate of zinc in dilute sulphuric acid (about 2–3M), and measure the potential across the plates with a potentiometer or high-resistance voltmeter. Then join a resistance of about 50 ohms across the electrodes and read the voltmeter every minute. Plot a graph of voltage against time. Repeat the experiment and after it has been going a few minutes, brush the hydrogen off the copper electrode and note the change in the voltage of the cell.

Expt. 11b-3

Set up a Daniell cell, consisting of a plate of copper in M/2-copper(II) sulphate solution separated by a porous partition from a zinc plate in M/2-zinc sulphate solution. Measure the potential difference on open circuit with a potentiometer or high-resistance voltmeter. Join a resistance of about 50 ohms across the electrodes and read the voltmeter every minute.

Next join the electrodes by a copper wire, note the potential difference again and measure it every 5 minutes. The following are the results of a typical experiment:

Time after short-circuiting (min)	0	5	10	15	20	25	30
Potential difference (V)	1·106	0·142	0·128	0·123	0·120	0·116	0·114

The first apparatus of the kind described in Expt. 11b-1 was made by Volta in 1800. Volta separated a plate of copper and a plate of zinc by a piece of cloth soaked in sulphuric acid, and found that when the plates were joined by a wire, an electric current flowed through it. A number of such units placed in series is known as the *voltaic pile*. This invention was of fundamental importance to the growth of the sciences of chemistry and electricity, for it was the first source of continuous electric current. Indeed, it was the only source of current electricity available for some years. As an example of its importance in the history of chemistry, we may note that a voltaic pile was used by Davy in his discovery of the alkali metals in 1807.

In Expt. 11b-3, as current is drawn from the cell the potential difference decreases. This is due to the formation of a layer of bubbles of hydrogen on the copper. The electrode potential of the copper is thus altered, in that it becomes more or less a plate of hydrogen. This effect of the hydrogen bubbles is called 'polarization'. It is, however, only one way in which polarization can occur, and a more complete discussion of polarization phenomena will be found below.

Electromotive force

Chemical action is taking place all the time in Volta's cell, but in the Daniell cell described in Expt. 11b-3, no action occurs until the copper and zinc are joined by a wire. Then, since zinc has a high negative electrode potential and copper a positive one, the zinc dissolves as zinc ions, and copper ions are deposited as copper:

$$Zn \rightarrow Zn^{2+} + 2e$$
$$Cu^{2+} + 2e \rightarrow Cu.$$

A current flows through the wire, negative electrons passing along it from the zinc to the copper. We may represent the change occurring when the cell is generating current by

$$Zn + Cu^{2+} \rightarrow Zn^{2+} + Cu.$$

It is clear from Expt. 11b-3 that the potential difference between the electrodes is a maximum when no current is being taken from the cell and the change

$$Zn + Cu^{2+} \rightarrow Zn^{2+} + Cu$$

is not occurring. The potential difference on open circuit is known as the 'electromotive force' (e.m.f.) of the cell. When the electrodes are joined by an external resistance, the potential difference between them falls; this is because the cell itself has a resistance, known as its *internal resistance*, through which its electromotive force has to pump the current.

When the chemical reaction

$$Zn + Cu^{2+} \rightarrow Zn^{2+} + Cu$$

occurs in a test-tube, i.e. when a piece of zinc is put into copper sulphate solution, heat is evolved (see Expt. 10a-5). However, when this reaction occurs in a cell, instead of thermal energy, *electrical* energy is produced. We will now discuss whether the electrical energy produced is equivalent in quantity to the heat evolved in the calorimeter. When current is taken from the cell, not all the energy produced by the chemical change is available for external use, because some is dissipated as heat in driving the current through the cell itself. If an attempt is made to measure the energy produced, for example, by the deposition of a mole of copper in the cell, even if the energy is withdrawn slowly as a very small current, the energy obtained will be different from that evolved in the calorimetric experiment.

Now instead of considering the possibility of withdrawing energy from the cell, let us consider what happens when a potential difference greater than the voltage of the cell is applied across the electrodes so as to oppose the generation of current by the cell. The chemical change is made to proceed in the opposite direction, i.e. copper dissolves and zinc is deposited. The external voltage need only be infinitesimally greater than the e.m.f. of the cell, hence the energy put in to reverse the chemical change in the cell equals that given out in its direct operation, provided we consider that only infinitesimally small currents are used in making the energy measurements. A cell of which this is true is called a 'reversible cell'.

If E^0 is the e.m.f. of the cell on open circuit, the electrical energy that could be generated by the deposition of 1 mole of copper (symbolized by ΔG^0), if the process could be carried out infinitely slowly, is nE^0F, where n is the valency of copper and F the charge on a gram-equivalent, i.e. 96,540 coulombs. For the standard Daniell cell

$$\Delta G^0 = - nE^0F$$
$$= - \frac{2 \times 1{\cdot}1 \times 96,540}{4{\cdot}18} = - 50,700 \text{ cal per mole.}$$

The heat evolved in the calorimeter, ΔH^0, when 1 mole of copper is formed, is $-50,100$ cals per mole. In this case, therefore, the energies liberated when the reaction occurs in these two ways are nearly the same. However, this is not so in most cells; ΔG^0 may be either greater or less than the heat of reaction, e.g. in the silver–copper cell, ΔG^0 is about 21,000 cals and ΔH^0 is about 30,000 cals per mole for the reaction

$$Cu + 2Ag^+ \rightarrow Cu^{2+} + 2Ag.$$

For further discussion, see p. 235.

Polarization

Polarization is defined as any change produced at an electrode by electrolysis or other means which causes its potential to differ from its reversible value.

One cause of polarization has already been referred to. In Expt. 11b-2, it was found that when a finite current was taken from Volta's cell, the potential difference between the electrodes fell owing to the accumulation of hydrogen on the copper electrode. Volta's cell is irreversible, since by applying an external voltage slightly greater than that of the cell, we do not reverse the chemical change occurring in the cell when it is operating in the usual way; and this type of polarization is found to occur in all irreversible cells.

In the similar experiment with the normal Daniell cell, which is reversible and therefore not subject to the above type of polarization, polarization was nevertheless found to occur slowly. This is due to the changes in concentration of the ions around the electrode due to deposition of copper and solution of zinc, and is known as concentration polarization.

Standard cells

A standard cell is a stable arrangement designed to give a constant e.m.f. which will not deteriorate with time. The normal Daniell cell was formerly used as a standard, but it is not in equilibrium because slow

EBC—Q

mixing of the electrolytes by diffusion through the partition occurs. A standard cell should fulfil the following conditions: it must be in equilibrium, it must contain chemicals which can be obtained in a state of purity, and, although it may suffer temporary changes when used, it must revert to its initial e.m.f. on standing. The cells most widely

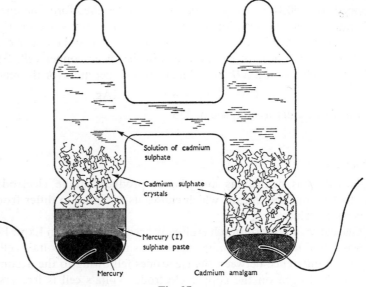

Solution of cadmium sulphate

Cadmium sulphate crystals

Mercury (I) sulphate paste

Mercury   Cadmium amalgam

Fig. 97

used now are the Weston and the Clark cells. The Weston cell, illustrated in Fig. 97, gives an e.m.f. of 1·0183 V at 25°C, and has only a very small temperature coefficient.

RELATIVE ELECTRODE POTENTIALS

The potential difference between an electrode and an electrolyte cannot be measured without the use of a second electrode to complete the circuit. In order to get information on the relative magnitudes of single electrode potentials it is therefore necessary to measure the e.m.f. of cells composed of a standard material for one electrode coupled with other materials in turn as the second electrode. The 'reference electrode' used is the molar hydrogen electrode (see below). The value of an electrode potential depends not only on the solution pressure of the electrode material, but also on the concentration of the ions in the electrolyte. It is therefore necessary to specify the latter, and it is usual to refer to molar solutions, i.e. those containing 1 mole of the ion per litre.

The electrode potential of various metals immersed in molar solutions of their ions relative to the molar hydrogen electrode are shown in the table.

STANDARD ELECTRODE POTENTIALS IN VOLTS AT 25°C

Lithium	Li^+	+2·96	Nickel	Ni^{2+}	+0·25
Potassium	K^+	+2·92	Tin	Sn^{2+}	+0·14
Sodium	Na^+	+2·71	Lead	Pb^{2+}	+0·13
Magnesium	Mg^{2+}	+2·37	Hydrogen	H^+	0·00
Aluminium	Al^{3+}	+1·66	Copper	Cu^{2+}	−0·34
Zinc	Zn^{2+}	+0·76	Silver	Ag^+	−0·80
Iron	Fe^{2+}	+0·44	Mercury	Hg^{2+}	−0·80
Cadmium	Cd^{2+}	+0·40	Gold	Au^+	−1·5

The manner in which such measurements are made will now be outlined.

The hydrogen electrode consists of a piece of platinum foil coated with platinum black (see p. 245) over which a stream of bubbles of hydrogen is maintained and which is immersed in a solution of unit hydrogen ion concentration. The arrangement behaves like a sheet of hydrogen (see Expt. 11b-5).

One electrode immersed in an electrolyte is known as a 'half-cell'. When two half-cells are put into electrical communication by means of a central compartment or bridge, usually containing potassium chloride solution, they form a voltaic cell. The e.m.f. of this cell is measured by means of a potentiometer capable of reading to thousandths of a volt. This e.m.f. is made up of the two potential differences at the two electrode-solution junctions:

$$E = e_H + e_M.$$

If e_H is taken as zero, then the measured e.m.f. gives the electrode potential of the metal relative to hydrogen (see table above).

There are technical difficulties in making such measurements, but once the electrode potentials have been obtained certain simplifications are possible. Thus the hydrogen electrode may be substituted by a more permanent and more easily handled electrode, such as the saturated calomel electrode. This electrode consists of the metal mercury covered by a layer of mercury(I) chloride in contact with a saturated solution of potassium chloride and gives a steady e.m.f. of 0·242 V at 25°C against the molar hydrogen electrode. If the calomel electrode is used as the reference electrode in measuring the electrode potential of a metal, say zinc, the latter can be obtained relative to hydrogen by subtracting 0·242 V from the measured e.m.f. In the case of zinc in contact with a normal solution of zinc ions, this is found to be 1·004 V; hence the electrode potential of zinc relative to the molar hydrogen electrode is 0·762 V.

Expt. 11b-4   Measurement of the e.m.f. of cells

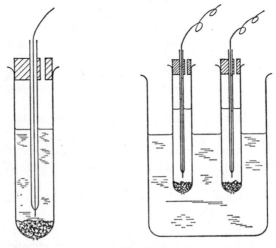

Fig. 98

A potentiometer reading to a thousandth of a volt will be required. This could consist of a 5-metre bridge wire and a high-resistance galvanometer, or equivalent instrument. A standard Weston cell and double pole double throw switch will also be needed. Alternatively, a valve voltmeter reading to 0·001 V can be used.

Each cell will consist of two half-cells, which are conveniently made as follows. A hole is blown in the end of a $5 \times \frac{5}{8}$-in. test-tube and the electrode material, in the form of a metal rod or wire supported by a piece of 5 mm o.d. glass tubing, is held by a split cork inserted in the mouth of the test-tube. The bottom of the test-tube is firmly packed to a depth of about 1 cm with small pieces of filter paper soaked in saturated potassium chloride solution. The cell is constructed by dipping two half-cells in a small beaker containing saturated potassium chloride solution. Fresh filter paper plugs should be used whenever the electrolyte being used in the half-cell is changed.

Use this equipment to measure the e.m.f. of cells, composed of pairs of the following electrodes materials in 0·1 molar solutions of their salts:

1. Copper in copper(II) sulphate.
2. Zinc in zinc sulphate.
3. Silver in silver nitrate,
4. Iron in iron(II) sulphate.
5. Nickel in nickel ammonium sulphate.
6. Lead in lead nitrate.
7. Magnesium in magnesium sulphate.
8. Cadmium in cadmium chloride or sulphate.

If a high resistance voltmeter is not available, and the potentiometer method is being used, first connect a 2 V accumulator across the potentiometer wire, then balance the e.m.f. of the standard cell and note the potentiometer reading. Substitute the experimental cell for the standard cell and again read the potentiometer. Calculate the e.m.f. of the cell from the known e.m.f. of the standard cell.

Preparation of a saturated calomel electrode

Put some purified mercury in a test-tube with a side arm to a depth of about 1 cm. Fit the test-tube with a rubber bung carrying a platinum electrode (made by sealing a short piece of platinum wire into a glass tube, and making contact with a little mercury). Above the mercury in the test-tube introduce, by means of a funnel, a paste made by grinding together mercury(I) chloride, mercury and water. The mercury layer should be covered by about 1 cm of paste. On to this run a slush of potassium chloride crystals, and fill up the test tube and side arm with a saturated solution of potassium chloride. Use the purest chemicals available, and avoid mixing the layers as far as possible.

In order to calculate single electrode potentials, it is convenient to measure the e.m.f. of a cell composed of the electrode as one half-cell and the calomel electrode as the other.

By subtracting 0·242 V (the potential of the saturated calomel electrode relative to the normal hydrogen electrode), the electrode potential of the metal in contact with 0·1 M solution of its salt is obtained.

In this way, find the electrode potentials of say, copper and zinc, and see whether the values add up to that of the copper-zinc cell, i.e. 1·1 volts.

Expt. 11b-5 Preparation and use of the hydrogen electrode

Hydrogen

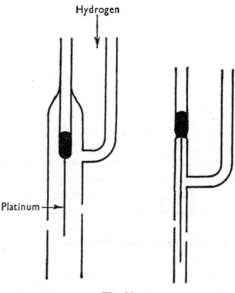

Platinum

Fig. 99

A simple hydrogen electrode may be made by sealing a platinum wire about 5 cm long into a soft glass tube as shown in Fig. 99. Deposit platinum black on the wire as described on p. 245. Wash the electrode well and arrange to pass a rapid stream of hydrogen over it. Use hydrogen from a cylinder or generate it by the action of dilute sulphuric acid on a zinc-copper couple, and wash the gas by passing it through a bubbler containing dilute caustic soda solution followed by one containing distilled water.

To set up a hydrogen electrode, immerse the electrode in a solution of hydrochloric acid containing a hydrogen-ion concentration of 1 gram-ion per litre. Make contact with the wire through a little mercury placed in the upper part of the tube. Couple the electrode with other half-cells (e.g. the calomel electrode, the copper and zinc electrodes), measure the e.m.f. and calculate the relative electrode potentials.

Redox potentials

Expt. 11b-6

Set up a half-cell consisting of a platinum electrode (a short length of platinum wire) immersed in an approximately molar solution of iron(II) sulphate in which some of the iron has become oxidized to iron(III). Couple this with a calomel electrode and measure the e.m.f. as described in Expt. 11b-4.

Repeat the measurement using half-cells containing approximately molar solutions of (i) Ce^{3+}/Ce^{4+}, (ii) Sn^{2+}/Sn^{4+}, (iii) $Cr_2O_7{}^{2-}/Cr^{3+}$, (iv) $MnO_4{}^-/Mn^{2+}$, (v) I^-/I_2, (vi) V^{2+}/V^{3+}.

Reactions at electrodes, where electrons are transferred, are oxidation-reduction reactions. For example, at the zinc electrode in the Daniell cell

$$Zn \longrightarrow Zn^{2+} + 2e,$$

which is an oxidation, and at the copper electrode

$$Cu^{2+} + 2e \longrightarrow Cu,$$

which is a reduction. The electrode potentials of the metals measured in earlier experiments refer to the system M/M^{n+}. The above experiment shows that processes in which a metal ion is being oxidized or reduced to a higher or lower oxidation state also give rise to electrode potentials, e.g. M^{2+}/M^+. For any oxidation/reduction system there is a corresponding potential, known as the redox potential. Some values of redox potentials are shown in the table.

Half-cell reaction	Redox Potential (volts)
$H_2O_2 + 2H^+ + 2e \longrightarrow 2H_2O$	+1·77
$MnO_4{}^- + 8H^+ + 5e \longrightarrow Mn^{2+} + 4H_2O$	+1·51
$Cr_2O_7{}^{2-} + 14H^+ + 6e \longrightarrow 2Cr^{3+} + 7H_2O$	+1·33
$V(OH)_4{}^+ + e \longrightarrow (VO_2)^{2+} + 2H_2O$	+1·00
$Fe^{3+} + e \longrightarrow Fe^{2+}$	+0·77
$I_2 + 2e \longrightarrow 2I^-$	+0·53
$(VO_2)^{2+} + e \longrightarrow V^{3+}$	+0·31
$Sn^4 + 2e \longrightarrow Sn^{2+}$	+0·15
$V^{3+} + e \longrightarrow V^{2+}$	−0·20

The redox potential is a measure of the oxidizing power of a system and enables us to predict whether a given oxidation-reduction process will occur. Try to predict whether solutions of

(i) Iron(III) sulphate will oxidize potassium iodide solution

(ii) Vanadium(II) sulphate will reduce a solution of iodine

(iii) Vanadium(II) sulphate will reduce a solution of iron(III) sulphate

(iv) Tin(IV) chloride will oxidize vanadium(II) sulphate solution.

Expt. 11b-7

(1) Carry out test-tube reactions to test the predictions made in connection with (i) to (iv) in the preceding paragraph. Note that V^{2+}(II) is pale lilac, V^{3+}(III) is green, VO_2^{2+}(IV) is blue, and $V(OH)_4^+$ (V) is orange.

(2) Mix solutions of iron(III) sulphate and vanadium(II) chloride. What colour change do you observe? Is this what you might expect from the values of the redox potentials?

(3) To a blue solution containing VO_2^{2+} ions, add a drop or two of dilute hydrogen peroxide solution. Does the colour change to orange, $V(OH)_4^+$?

CONCENTRATION CELLS

A voltaic cell consisting of two half-cells containing the same electrode material but different concentrations of the ions in solution, generates an e.m.f., for, although the electrolytic solution pressure is the same in each cell, the tendency for ions to deposit on the electrodes is different. From the measured e.m.f. of the cell, the relative concentrations of the solutions can be calculated. The hydrogen concentration cell is of special interest, as it affords a means of measuring the pH of a solution. The cell consists of two half-cells containing hydrogen electrodes; one contains a solution molar in hydrogen ions, the other contains the solution whose pH is required. The e.m.f. of the cell is measured, and from this the pH can be calculated, as shown below. A simplification can be introduced into the experimental arrangement by the use of a metal electrode instead of the hydrogen electrode in the unknown solution, and of a calomel electrode as reference instead of the hydrogen electrode. The metal antimony has been shown to give a series of e.m.f. in solutions of a wide pH range, which bear a linear relation to those obtained with a hydrogen electrode. It is therefore suitable for the electrode in the solution of unknown pH. See p. 237.

It can be shown that the potential difference, e, between a metal of electrolytic solution pressure, P, and a solution of its ions (valency n)

of bombardment pressure, p, is given by $e = \dfrac{RT}{nF} \log_e \dfrac{P}{p}$, where R is the gas constant, T the absolute temperature and F is 96,540 coulombs. It follows that the e.m.f. of a concentration cell is

$$E = \frac{RT}{nF} \log_e \frac{P}{p_2} - \frac{RT}{nF} \log_e \frac{P}{p_1}$$

i.e.
$$E = \frac{RT}{nF} \log_e \frac{p_1}{p_2}$$

$$= \frac{RT}{nF} \log_e \frac{c_1}{c_2}$$

where c_1 and c_2 are the ionic concentrations in the two half-cells, assumed to be proportional to the bombardment pressures of the solutions. If the measurements are made at room temperature, this e.m.f. is approximately equal to $\dfrac{0.059}{n} \log_{10} \dfrac{c_1}{c_2}$ volts. For example, if $n = 1$, and $c_1 = 10c_2$, then $E = 0.059$ volts.

Expt. 11b-8 Measurement of the e.m.f. of copper concentration cells

The apparatus described in Expt. 11b-4 will be required. Prepare about 100 cm³ each of 0·5M, 0·05M, and 0·005M solutions of copper sulphate. Measure the e.m.f. of cells consisting of a saturated calomel electrode and a copper electrode containing each copper sulphate solution in turn. Use a fresh plug of filter paper in the cell for each measurement.

Specimen results

0·5M-$CuSO_4$ — calomel	0·062 V
0·05M-$CuSO_4$ — calomel	0·044 V
0·005M-$CuSO_4$ — calomel	0·021 V

from which by subtraction,

0·5M-$CuSO_4$ — 0·05M-$CuSO_4$	0·018 V
0·05M-$CuSO_4$ — 0·005M-$CuSO_4$	0·023 V
0·5M-$CuSO_4$ — 0·005M-$CuSO_4$	0·041 V

In a semi-molar solution of copper sulphate the 'bombardment pressure' due to the copper ions is less than would be obtained from an ideal solution because of the attractive forces between these and the sulphate ions. The effective concentration of the copper ions (the 'activities') is not 0·5M but is found to be only 0·07M. In 0·05M and 0·005M solutions, the effective concentrations are respectively 0·019M and 0·003M. If these values are inserted in the relation

$$E = \frac{0.059}{n} \log_{10} \frac{c_1}{c_2}$$

we find the following theoretical values for the e.m.f. of the concentration cells at room temperature: 0·016 V, 0·023 V and 0·039 V, with which the above specimen results are in fair agreement.

Expt. 11b-9   Iron/cadmium concentration cells

By measuring the e.m.f. of cells consisting of a calomel half-cell and cadmium or iron half-cells, find the electrode potentials of iron and cadmium in solutions of their salts of molarities varying from M through 0·1M to 0·01M. Plot a graph showing how the electrode potential of each metal varies with the concentration of its ions in the half-cells.

This is an example of a concentration cell in which the electrodes are of different metals. Reference to the table of standard electrode potentials would suggest that the reaction taking place is represented by

$$Fe + Cd^{2+} \longrightarrow Fe^{2+} + Cd.$$

As the reaction occurs the concentration of iron ions increases and that of the cadmium ions decreases. What information do the graphs obtained above give about the course of the reaction in which the initial concentrations were 0·1M? Study the following table:

ELECTRODE POTENTIALS IN VOLTS OF IRON AND CADMIUM IN SOLUTIONS OF VARIOUS CONCENTRATIONS OF THEIR IONS

	Iron	Cadmium
1M	+0·44	+0·40
0·1M	+0·47	+0·43
0·01M	+0·50	+0·46
0·001M	+0·53	+0·49

It follows from the above that the e.m.f. of a cell made up of half-cells containing 0·1M solutions of both ions would be +0·43 − (+0·47) = −0·03 V. Consider what would happen if it were possible for the cadmium ion concentration to fall (as a result of reaction occurring in the cell) from 0·1M to 0·01M. This would involve the deposition from 1 litre of solution of 0·9 moles of cadmium as metal. This means that

0·9 moles of iron would have been dissolved from the electrode and entered the solution as iron(II) ions, whose concentration would then become $0·1 + 0·9 = 1·0$M. The table shows that the e.m.f. of the cell would then have become $+0·46 - (+0·44) = +0·02$. The sign of the e.m.f. has changed showing that the cell reaction has been reversed.

In practice, of course, the e.m.f. of the cell would fall to zero and the reaction reach equilibrium:

$$Cd + Fe^{2+} \leftrightharpoons Cd^{2+} + Fe.$$

This example of a reversible reaction illustrates several important points: it shows that if the electrode potentials of two metals are very close to each other the interaction between one metal and the ions of the other will be reversible.

It also follows that the e.m.f. of the cell made up from half-cells consisting of the two metals in equimolar solutions of their ions will be a measure of the position of the equilibrium between the forward and back reactions. If the e.m.f. is small, the equilibrium will be such that the concentrations of the two ions are of the same order of magnitude. However, if the equilibrium is well over to one side, the e.m.f. of the cell will be correspondingly large. Thus, compare the e.m.f. of the standard Cd/Fe cell, 0·04 V, with that of the standard Daniell cell, Cu/Zn, 1·1 V. The equilibrium position of the reaction

$$Zn + Cu^{2+} = Cu + Zn^{2+}$$

is almost completely over to the right-hand side.

In fact there is a connection between the standard e.m.f. of a cell and the equilibrium constant, K, for the reaction occurring in the cell.

If e^o represents the electrode potential of a metal immersed in a solution of its ions under standard conditions (i.e. $c = 1$) then the electrode potential in a solution of ion concentration c_1 is given by

$$e = e^o + \frac{RT}{nF} \log_e c_1$$

In a cell consisting of two half-cells made from two different metals, M_1 and M_2,

$$E = E^o + \frac{RT}{nF} \log_e \frac{c_{M_1}}{c_{M_2}}$$

where $E^o = e^o$ for $M_1 + e^o$ for M_2, and c_{M_1} and c_{M_2} represent the concentrations of the metal ions in the two half-cells and E is the e.m.f. of the cell on open circuit.

When the cell reaction is at the equilibrium point, $E = 0$ and

$$E^0 = -\frac{RT}{nF}\log_e\frac{c_{M_1}}{c_{M_2}},$$

where c_{M_1} and c_{M_2} are equilibrium values.

The energy supplied by the cell is nE^0F, symbolized by ΔG^0,

so
$$\Delta G^0 = -RT\log_e\frac{c_{M_1}}{c_{M_2}}.$$

For such a reaction, K, the equilibrium constant, $= \frac{c_{M_1}}{c_{M_2}}$,

so
$$\Delta G^0 = -RT\log_e K.$$

As a reaction proceeds, the energy available from the reactants (the sum of their 'free energies') falls. In a reversible reaction, this will be true of both the forward and back reaction. If we think of the equilibrium position being approached from either side, it is clear that when the reaction is in equilibrium, the free energies of the reactants and products are at a minimum and $\triangle G$ for the reaction is zero. To avoid confusion, we must distinguish between G^0, the standard free energy of a substance, i.e. its free energy in the standard state, and G, its free energy under any conditions, a quantity whose value may vary with temperature and concentration. The algebraic sum of the values of G for the reactants and products is zero when they are in equilibrium with each other, but we can still calculate a value for ΔG^0 from the values of G^0 for the reactants and products. It is this value, ΔG^0, which is related to the equilibrium constant and which gives information about the position of the equilibrium.

MEASUREMENT OF pH

Expt. 11b-10   Hydrogen concentration cell

Place the hydrogen electrode described on page 230 in an M/10-hydrochloric acid solution, couple it with the calomel electrode and measure the e.m.f. of the cell. Then replace the M/10 with an M/100-hydrochloric acid and repeat the measurement. The difference between the e.m.f. of the two cells should be equal to the e.m.f. of a hydrogen concentration cell in which $c_1/c_2 = 10$, which is about 0·059 volts.

Expt. 11b-11 Measurement of the pH of a solution

Use the calomel electrode as reference, and place a rod of antimony in the solution whose hydrogen-ion concentration is to be measured. The antimony rod should be cleaned with emery paper, left in the air for a minute or two to form an oxide film on the surface, and washed with distilled water. It should be washed between each measurement, and 3 or 4 minutes should be allowed before the reading is taken after immersing the rod in the solution. From the measured e.m.f. of the cell, $Sb/Sb_2O_3/H^+soln/KCl/saturated$ calomel, the pH of the solution under test can be obtained from the table given below. Measure the pH of (1) distilled water, (2) tap water, (3) a buffer solution of known pH (prepared as described in Expt. 12b-13), (4) any liquid of unknown pH, e.g. vinegar, port wine, milk at daily intervals, etc.

E.m.f. of antimony-saturated calomel cell	pH	E.m.f. of antimony-saturated calomel cell	pH
0·119	2·5	0·382	7·5
0·145	3·0	0·407	8·0
0·172	3·5	0·430	8·5
0·199	4·0	0·456	9·0
0·226	4·5	0·482	9·5
0·258	5·0	0·509	10·0
0·282	5·5	0·535	10·5
0·307	6·0	0·568	11·0
0·332	6·5	0·587	11·5
0·357	7·0		

(adapted from Britton and Robinson, *J. Chem. Soc.* 1931, p. 466)

Specimen results

The measured e.m.f. of the cell, $Sb/Sb_2O_3/distilled\ water/KCl/satur$-ated calomel was 0·352 V. The table shows the values of the e.m.f. of the cell, $Sb/Sb_2O_3/solution/KCl/saturated\ calomel$, and the corresponding pH values. Hence the value obtained for the pH of distilled water in this experiment was 6·98.

Other results were

    (1) tap water   e.m.f. 0·429; pH 8·47
    (2) buffer solution, pH 2·2, e.m.f. 0·100; pH 2·20
    (3) sherry wine,   e.m.f. 0·188; pH 3·88.

Expt. 11b-12 Potentiometric titrations

Electrometric methods can be used to follow the course of a reaction in which the pH is changing. The end-point of an acid-alkali titration can be determined by following the changing pH of one solution while the other is being added, using the method of Expt. 11b-11. An abrupt change in e.m.f. occurs at the equivalence point. Typical curves showing the changes in pH as the titration proceeds are shown in the figures. The method is useful in titrating coloured liquids where the colour of an indicator would be obscured, and has also been widely used in the study of complex acids.

(i) Find the end-point in a titration of decimolar caustic soda with decimolar hydrochloric acid. Put the antimony electrode described in Expt. 11c-8 in 25 cm³ of the alkali, run in 23 cm³ of the acid and measure the e.m.f. against the calomel electrode. Then run in the acid 0·2 cm³ at a time and repeat the e.m.f. measurement after each addition. Continue until 27 cm³ of the acid have been added. Plot a graph of cm³ of acid added against the potentiometer readings, and note where the equivalence point occurs.

Specimen result

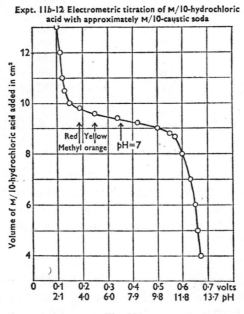

Expt. 11b-12 Electrometric titration of M/10-hydrochloric acid with approximately M/10-caustic soda

Fig. 100

Fig. 100 shows a curve that was obtained for the titration of 10 cm³ of M/10-hydrochloric acid with approximately M/10-caustic soda. Observations were also made on the colour of the methyl orange as the titration proceeded. The curve shows that (1) the equivalence point occurs at a pH of almost exactly 7, (2) the change point of the methyl orange is at pH 4·2, (3) 9·4 cm³ of alkali are equivalent to 10 cm³ of acid, (4) the methyl orange indicates that 9·6 cm³ of alkali are required, a result that differs by about 2% from the true equivalence value.

Fig. 101 shows the curve for the titration of M/10-sodium carbonate solution by approximately M/20 sulphuric acid, and shows the sharp change in pH that occurs when the carbonate is completely decomposed and the rather less sharp change at the bicarbonate stage at pH 8·4.

Specimen results

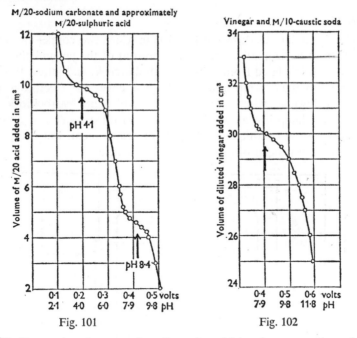

Fig. 101

Fig. 102

(ii) Determine the quantity of acetic acid in vinegar or the acidity of sour milk by this method.

Fig. 102 shows the curve for the titration of vinegar (diluted ten times) by M/10-caustic soda, and gives the acidity of the vinegar as M/3.

Simplified methods have been adopted for determining the end-point in electrometric titrations. Two suitable metals are immersed in the solution to be titrated and connected to a high-resistance millivolt-meter. Readings of the millivoltmeter are taken as the standard solution is run in, and are plotted against the number of cm³ added. The point of inflexion in the graph obtained indicates the end-point of the titration. Experiment with the method by using (*a*) antimony and tungsten electrodes for an acid-alkali titration, (*b*) platinum and tungsten electrodes for an iodine-thiosulphate titration. Examine the behaviour of other pairs of electrodes, e.g. copper and platinum, graphite and copper, etc.

(c) Specific and equivalent conductance

Expt. 11c-1

Use the apparatus shown in Fig. 103 in which two copper electrodes are fixed horizontally, one on the bottom of the jar and the other above it, insulated from it by a small flat cork, and joined in series with a battery, lamp and switch. Just cover the electrodes with distilled water, switch on and add drops of dilute hydrochloric acid until the lamp just glows. Stir in more distilled water and note the effect on the conductance. Repeat the experiment, but use glacial acetic acid instead of dilute hydrochloric acid.

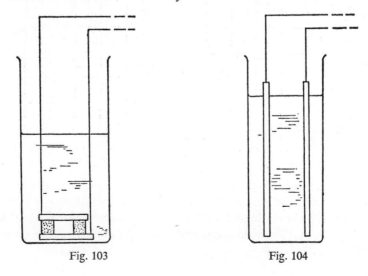

Fig. 103   Fig. 104

Expt. 11c-2

Rearrange the electrodes vertically (see Fig. 104). Put 1 cm of distilled water at the bottom of the jar, and add drops of dilute hydrochloric acid until the lamp just glows. Add more distilled water and notice that the glow of the lamp is unchanged. Now repeat the experiment using glacial acetic acid instead of hydrochloric acid. Note that dilution increases the conductance as measured in this manner.

In Expt. 11c-1 the specific resistances of the solutions are compared. (The specific resistance, σ, of a material is the resistance between opposite faces of a centimetre cube.) For solutions, it is more convenient to use the inverse of the specific resistance, namely, the 'specific conductance', k. Thus, $k = 1/\sigma$. The experiment shows that the specific conductance of the electrolyte decreases with increasing dilution. If the dilution were continued indefinitely until the solution became indistinguishable from distilled water, the conductance would fall to a very low value. Kohlrausch, in 1894, measured the conductance of redistilled water and found that, as he was able to make progressively purer water, the conductance fell to a steady value, below which it did not go however carefully the water was purified. It is thus evident that water itself has a slight conductance. The value found by Kohlrausch at 18°C was 0.40×10^{-6} reciprocal ohms (mhos).

The decrease of conductance on dilution is interpreted quite simply on the ionic theory as follows. Dilution disperses the solute ions in a larger volume of solution and hence there are fewer ions between the electrodes to 'ferry' the charges across. The large difference in the specific conductances of acetic and hydrochloric acids is notable: the experiment showed that the value for acetic acid was much less than that for hydrochloric acid. It follows from the ionic theory that there must have been fewer ions between the electrodes in the solution of acetic acid. Now the concentration of this acid in moles per litre was greater than that of the hydrochloric acid. This means that only a fraction of acetic acid molecules exist as ions in the solution, and indicates the possibility of a substance existing in both ionized and un-ionized forms in solution.

Considerable interest has centred around the question of how such molecules as those of acetic acid become ionized. Faraday's view was that, in the case of 'ionic' substances, the ions exist in the solid state, and that is the view held today. In order to explain the formation of conducting solutions when non-ionic substances such as acetic acid (which is a non-conductor when pure) are dissolved in water, Clausius assumed (1857) that a fraction of the molecules dissociate into ions on going into solution. Arrhenius developed this idea and assumed that an equilibrium exists in the solution between the ions and the un-ionized molecules (1883). Thus, in a solution of acetic acid, a fraction of the molecules would exist as ions and impart a certain conductance to the solution. Expt. 11d-2 throws more light on this theory of ionic dissociation.

In Expt. 11c-2 the electrodes are arranged so that, on dilution, most of the solution is still included between them. Dilution of the hydrochloric acid produces no change in the conductance as measured in this way. This is easily understood in terms of the ionic theory, for there are as many ions 'plying as ferries' carrying the charges from one electrode to the other after dilution as there were before—they are merely spread out to cover a larger area of electrode. The behaviour of acetic acid is different, in that the conductance *increases* with dilution (see Fig. 105). In terms of the ionic theory, this is capable of two possible interpretations: (i) there are more ions to conduct the current from one electrode to the other in the more dilute solution; (ii) the ions move faster in the diluter solution, hence more charge is carried across per second, giving an increased conductance. The first interpretation was that developed by Arrhenius, and it undoubtedly holds for those substances, such as acetic acid, which are non-conductors until dissolved in an ionizing solvent such as water. The second interpretation contains the germ of the more recent theory of Debye and Hückel for strong electrolytes.

The role of the solvent should be noted. Water reacts with those substances, such as acetic acid, which are not conductors (and so do not consist of ions) to form a conducting solution. This is explained in terms of the production of ions thus:

$$HA + H_2O = H_3O^+ + A^-.$$

The equilibrium is well over to the right-hand side if the acid is hydrochloric acid, but in the case of acetic acid, an appreciable conversion to ions only takes place in very dilute solutions (see tables on pages 244 and 261).

In Expt. 11c-2, the conductance of a fixed mass of solute dispersed in a variable volume of solution was measured. It illustrates the meaning of the quantity known as the 'equivalent conductance', Λ_v, which

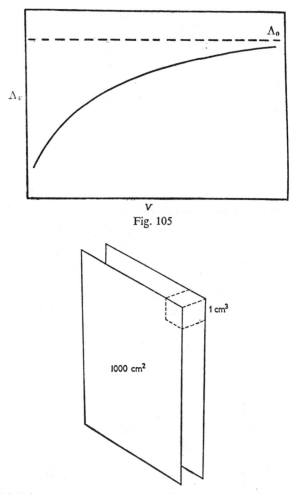

Fig. 105

Fig. 106. Prism containing 1 gram-equivalent of solute in v litres of water

is the total conductance of 1 gram-equivalent of solute dispersed in
v litres. The equivalent conductance of a given solution is obtained
from measurement of the specific conductance, k, by means of the
relation $\Lambda_v = kv \times 1000$ (see Fig. 106). Results of such determinations
show that the value of Λ_v for most solutes increases as the solution
is diluted, and approaches a maximum, Λ_0, known as the 'conductance
at zero concentration'. The values of Λ_v for various values of v are
shown below for a few common electrolytes.

Λ_v at 18°C, ohms^{-1} cm^2

v (litres)	NaCl	KCl	HCl	$\frac{1}{2}K_2SO_4$	$\frac{1}{2}CuSO_4$	CH$_3$COOH (at 25°C)
1	74·3	98·2	—	71·6	25·7	1·6
2	80·1	102·4	—	78·4	30·7	2·2
5	87·7	107·9	—	87·7	37·6	3·5
10	92·0	112·0	351·4	94·9	43·8	5·0
20	95·7	115·7	358·4	101·9	51·1	7·1
100	101·9	122·4	369·3	115·8	71·6	16·0
500	105·5	126·2	375·3	124·6	91·8	34·0
1000	106·4	127·3	375·9	126·9	98·4	47·6
∞	108·9	130·0	380·0	133·0	114·4	388

Expt. 11c-3   Conductance measurements

The conductance cell shown in Fig. 107 consists of two platinum electrodes e fixed in a cell made of resistance glass, in which is placed the solution whose conductance is to be measured. The cell is connected up in a bridge circuit similar to that of the Wheatstone bridge, so that its resistance may be measured. The Kohlrausch bridge differs from the Wheatstone bridge in that alternating instead of direct current is used, and headphones of low resistance instead

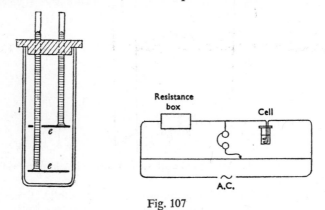

Fig. 107

of a galvanometer are used to detect the balance point. Alternatively, an a.c. meter or 'magic eye' may be used. If direct current were to be employed, electrolysis of the cell contents would occur and the conductance be altered. A metre bridge wire and a resistance box reading from 10 to 1000 ohms are required to complete the circuit.

To minimize the effect of any slight deposition on the electrodes from electrolysis, and to obtain a sharp null-point, the platinum must be coated with platinum black. To do this, clean the electrodes in hot concentrated nitric acid, wash well with distilled water, and immerse in a solution containing about 0·5 g of platinic chloride and 0·01 g of lead acetate in 50 cm³ of bench dilute hydrochloric acid. Connect to two accumulators in series for about 15 minutes and alter the direction of the current at frequent intervals. A fine deposit of platinum is formed on the electrodes. Wash them and continue the electrolysis for 5 minutes in a solution of dilute sulphuric acid, altering the direction of the current several times. Finally, wash well in distilled water.

It is important not to touch the electrodes or to disturb their relative position. They should not be removed from the stopper carrying them, and should always be washed by immersion in distilled water or in the solution in which they are going to be used.

For precision work it is necessary to use 'conductance water', but satisfactory results will be obtained in the experiments described below, if distilled water straight from the still is used. Its conductance may be further reduced by running it through a de-ionizing column. As the conductance of solutions has an appreciable temperature coefficient, surround the cell by a large vessel of water in order to keep the temperature constant during the experiment.

The cell constant. If a solution of known specific conductance, k, is placed in the cell, and the resistance, R, measured, a constant can be calculated which can be used thereafter to convert resistance measurements directly into specific conductances. If b is this 'cell constant', then $b = Rk$.

Prepare an M/50 solution from AnalaR potassium chloride. Wash the electrodes by immersion in some of this solution, then place enough solution in the cell to cover the electrodes to a depth of about 1 cm, and fix them in position. When determining the resistance, arrange the resistance of the box so that the balance point is obtained somewhere near the centre of the wire if possible. Make a second determination, using another value of the resistance box. Calculate the cell constant, using the appropriate value of k from the table:

SPECIFIC CONDUCTANCE OF M/50-POTASSIUM CHLORIDE SOLUTIONS

Temp. (°C)	10	18	25
k (ohms⁻¹ cm⁻¹)	0·00200	0·00240	0·00277

Conductance of copper(II) sulphate solutions

Prepare a half-molar solution of recrystallized (or AnalaR) copper(II) sulphate by dissolving 12·47 g in freshly distilled water and making up to 100 cm³. Use some to wash out the cell and the electrodes, taking care not to disturb their position. Then place some solution in the cell and measure the resistance as described above. Calculate the specific conductance from $K = b/R$. By successive dilutions, prepare M/4, M/8, M/16, . . ., M/512 solutions and measure the conductance of each. Calculate the conductances, $\Lambda_v = kv$, and plot Λ_v against v, the dilution, i.e. the volume, in cm³ containing 1 mole. Estimate the value to which Λ_v is approaching as the dilution becomes very great, i.e. estimate the value of Λ_0. See if these results are in agreement with the Arrhenius theory by plotting Λ_v/v against $1/\Lambda_v$ (see p. 262). The graph will be found not to be a straight line, showing that the behaviour of copper(II) sulphate solutions is not in accordance with the Arrhenius theory of electrolytic dissociation. However, the curve in Fig. 108 can be extrapolated and the value of Λ_0 at 18°C is seen to be about 118.

Specimen result

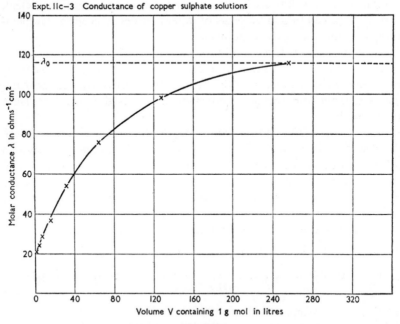

Expt. IIc-3   Conductance of copper sulphate solutions

Fig. 108

The behaviour of copper(II) sulphate should be contrasted with that of acetic acid, a weak electrolyte. At a dilution of $M/100$-copper sulphate is almost completely dissociated and the value of Λ_v is close to that of Λ_0. But in the case of acetic acid, Λ_v continues to increase on dilution and even at $M/1000$, Λ_0 is not approached. This shows why the value of Λ_0 for a weak electrolyte cannot be obtained by extrapolation. For further treatment of ionic dissociation see p. 259.

MIGRATION OF IONS

Expt. 11c-4

Fill a wide U-tube with an approximately 5% solution of copper(II) sulphate (see Fig. 109). Push a plug of cotton wool into each limb

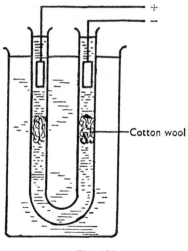

Cotton wool

Fig. 109

as shown. Place copper-foil electrodes in the solution and connect them to a dry battery of 50/100 V. Immerse the U-tube in a large bath of cold water and leave the current on for some minutes. The exact conditions will have to be determined. Note the increase in the depth of the colour of the solution around the anode and the decrease in that around the cathode.

Repeat the experiment using platinum electrodes, and compare the decreases in colour of the anode and cathode solutions.

Expt. 11c-5

(i) Fill the lower half of a U-tube with a warm 3% agar solution containing about 5% of copper(II) sulphate, and allow it to set. Pour on to the jelly some agar solution containing potassium nitrate. When this too has set, insert platinum electrodes near the top of the jellies and connect to a battery of about 50 volts. Note the movement of the blue boundary as the copper ions move towards the negative electrode.

(ii) If copper(II) chromate is used in this experiment, the movement of both ions is visible, the blue copper ions moving towards the cathode and the orange chromate ions towards the anode. Use a U-tube with a central tube attached (see p. 169). Precipitate some copper chromate by adding potassium chromate solution to copper(II) sulphate solution. Dissolve the brown precipitate in the least quantity of dilute nitric acid. Dissolve quite a lot of urea in this solution to increase its density. Support two platinum electrodes in the limbs of the U-tube and half-fill it with dilute nitric acid. Slowly run in the acidified solution of copper chromate from a funnel so that it forms a separate layer below the acid, which is pushed up until the electrodes are immersed. Connect the electrodes to a d.c. supply of about 30/50 V and keep the U-tube cool by immersion in a large beaker of cold water. Leave the voltage on for about 20 minutes until the coloured solutions of the two ions are clearly visible above the greenish solution of copper(II) chromate.

It was shown by Kohlrausch in 1876 that the conductance of a solution could be expressed as the sum of two parts, attributable to the cations and anions respectively. The contributions of the two ions are called the ionic conductances, and are represented by the symbols λ^+ for the cation and λ^- for the anion. This *Law of the Independent Migration of Ions* is based on results derived from the values of Λ_0 for various electrolytes, e.g.

$$\text{For} \quad \text{KCl:} \quad \Lambda_0 = \lambda_{0K^+} + \lambda_{0Cl^-} = 129 \cdot 1$$

$$\text{and for} \quad \text{KNO}_3\text{:} \quad \Lambda_0 = \lambda_{0K^+} + \lambda_{0NO_3^-} = 125 \cdot 5.$$

$$\text{For} \quad \text{NaCl:} \quad \Lambda_0 = \lambda_{0Na^+} + \lambda_{0Cl^-} = 108 \cdot 1$$

$$\text{and for} \quad \text{NaNO}_3\text{:} \quad \Lambda_0 = \lambda_{0Na^+} + \lambda_{0NO_3^-} = 104 \cdot 6.$$

The difference in Λ_0 for the chlorides is 21·0, and for the nitrates, 20·9.

The expected constancy of $(\lambda_{K^+} - \lambda_{0Na^+})$ is thus borne out. The figures also show that the conductances of the sodium and potassium ions differ appreciably, and this is confirmed by the results obtained by measuring the ionic conductances directly. In the case of coloured and certain other ions, this may be done by direct measurements of the movement of the boundaries.

ION CONDUCTANCES AND MOBILITIES

		Absolute velocity in cm per sec under unit potential gradient	
Ion	Conductance (18°C)	From conductance ($\times 10^{-4}$)	By direct measurement (Steele) ($\times 10^{-4}$)
H^+	318	32·9	28·0
K^+	64·7	6·7	5·5
Na^+	43·6	4·5	3·2
Li^+	33·4	3·5	1·9
Ag^+	54·0	5·6	—
OH^-	174	17·9	15·8
Cl^-	65·4	6·8	5·3
NO_3^-	61·8	6·4	—
CH_3COO^-	35	3·8	—
$\frac{1}{2}SO_4^{2-}$	68·5	7·1	3·0 (4·5)
$\frac{1}{2}Cu^{2+}$	45·9	4·8	1·8 (2·9)

Expt. 11c-6 Direct measurement of ion velocity

Make a solution containing 3 g of agar and 5 g of potassium nitrate in 100 cm³ of distilled water. Add enough bromthymol blue to make the solution distinctly green. Pour while hot into the inner tube of a small condenser, clamped vertically, and allow to set (see Fig. 110). Put a little dilute nitric acid at E and F, and keep the jelly cool with a good stream of cold water through the condenser. Insert platinum electrodes as shown and apply a d.c. voltage of 30–50 V from a dry battery. Within a few minutes coloured bands will appear, the yellow band showing the movement of hydrogen ions away from the anode and the blue band showing the movement of hydroxyl ions away from the cathode. The rate of migration of the two ions may be compared by noting the relative distances moved by the two colour fronts in, say, 10 or 20 minutes.

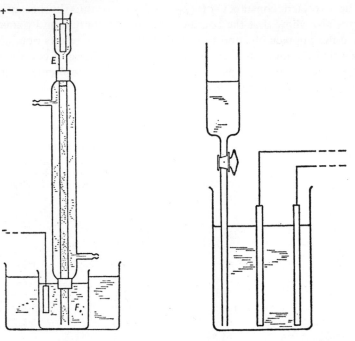

Fig. 110   Fig. 111

CONDUCTROMETRIC TITRATIONS

Expt. 11c-7

Assemble the apparatus described in Expt. 11c-2 (see Fig. 111).
Put about 5 cm of distilled water in the bottom of the cell and add
molar hydrochloric acid drop by drop until the lamp just glows.
Add the same number of drops of molar caustic soda solution to
a roughly equal volume of water and dissolve some urea in it to
increase its density. Place this solution in a tap-funnel and run it
very slowly into the cell so that it forms a separate layer below the
acid. Switch on the current, note the brightness of the lamp; then
stir up the solutions and note the decrease in conductance.

Expt. 11c-8

Place several hundred cm³ of distilled water in the apparatus used
above, and add dilute hydrochloric acid until the lamp just glows.

Place in a burette an approximately M/10 solution of caustic soda, and run it into the cell, stirring with a glass rod. Note that the lamp dims, and then gets brighter again.

In these two experiments the reaction occurring in the cell is

$$H^+ + Cl^- + Na^+ + OH^- \rightarrow Na^+ + Cl^- + H_2O.$$

Since the ionization of the water is negligibly small, the net result of the reaction is that hydrogen ions are replaced by sodium ions. The conductance of the latter is less than that of the hydrogen ions, so the lamp dims. Addition of caustic soda in excess of that required to neutralize the hydrochloric acid, produces additional sodium and hydroxyl ions and the conductance of the solution therefore rises again.

Conductometric titrations are chiefly of use in estimating the concentration of very dilute, coloured or turbid solutions where indicators cannot be used.

Expt. 11c-9

Add about 10 cm³ of M/20-barium hydroxide solution to distilled water in the apparatus used above and colour it with phenolphthalein. Run in M/20-sulphuric acid from a burette, stirring the solution continuously. Note that the lamp goes out at the end-point and that the indicator is simultaneously decolorized.

Expt. 11c-10   Conductometric titrations

Use the electrodes and Kohlrausch bridge described in Expt. 11c-3. Remove the electrodes from their cell and clamp them carefully near the bottom of a 100 cm³ beaker. Alternatively, a simpler system can be used by making use of electric light bulb leads as electrodes. Use a 4·5 V dry battery and a milliammeter shunted to read 0·1 mA as a measure of the conductance. A key completes the circuit.

Place 25 cm³ of approximately M/10-caustic soda solution in the beaker and fill a burette with M/20-sulphuric acid. Run 15 cm³ of the acid into the beaker and mix well, taking care not to disturb the electrodes. Note the conductance as measured by the bridge circuit or the milliammeter. Run in another cm³ of acid, stir, and again measure the conductance. Continue in this way until a total of about 35 cm³ of acid has been added. Plot a graph of the number of cm³ of acid added against the conductance measurement. A graph similar to that shown in Fig. 112 should be obtained. From it, calculate the concentration of the caustic soda, and check by titration, using an indicator in the usual manner.

Specimen result

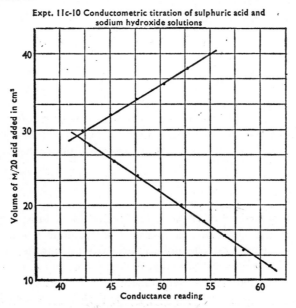

Expt. IIc-10 Conductometric titration of sulphuric acid and sodium hydroxide solutions

Fig. 112

Repeat the conductometric titration of caustic soda using M/10 solutions of (*a*) acetic acid, (*b*) *p*-nitrophenol, (*c*) phenol, and (*d*) periodic acid. Note the different forms of the graphs. From the results, can you decide which is the stronger acid, (*b*) or (*c*)?

12 Chemical equilibria

(a) Equilibrium in the gaseous state

Expt. 12a-1   The equilibrium $N_2O_4 \rightleftharpoons 2NO_2$

Fit a round-bottomed flask with a rubber bung carrying two delivery tubes, each fitted with a piece of rubber tubing and a clip. Attach one tube to a filter flask and thence to a water-pump. Pass nitrogen tetroxide, obtained by heating lead nitrate, through the flask until the colour is pale yellow. Close the clips and turn on the pump. When most of the gas has been removed from the filter flask, momentarily open the clip connecting the latter to the round-bottomed flask. At first the colour pales owing to the rarefaction of the gas. The next moment, the colour deepens, because increased dissociation increases the concentration of the coloured species, NO_2. Carry out several repetitions in order to find the best conditions for the particular apparatus used. With the clips closed, gently warm the flask and note the effect on the colour of the contents.

A number of reversible chemical reactions will be familiar. Some reactions which may commonly go in one direction, may with a change of conditions go in the reverse direction. If reactants and products are confined in a closed system, an equilibrium is established in which all are present in a certain proportion. This is a state of dynamic equilibrium in which both reactions are occurring at the same rate, i.e. the same quantity of reactant A is produced per minute by one reaction as is decomposed per minute by the reverse reaction.

Consider the following examples:

(i) $2HI \rightleftharpoons H_2 + I_2$,   (ii) $N_2 + 3H_2 \rightleftharpoons 2NH_3$.

Reaction (i) was first studied by Bodenstein (1899), who showed that, at a given temperature, the percentage of hydrogen iodide in the equilibrium mixture was the same whether the state of equilibrium had been reached by starting with hydrogen iodide, or with hydrogen and iodine in equimolecular proportions. Changes in pressure have no effect on the equilibrium position. In reaction (ii), the well-known industrial synthesis of ammonia, the percentage of ammonia in the equilibrium mixture is increased by an increase in the total pressure, but decreased by a rise of temperature.

Le Chatelier's principle may be applied to these chemical equilibria and may be used to predict the effect of changes in temperature and pressure on the equilibrium position. Consider first the effect of temperature.

In the reaction

$$A + B \rightleftharpoons C + D, \quad \Delta H = -Q$$

a rise in temperature will shift the equilibrium position in a direction leading to the absorption of heat, i.e. the percentage of C and D will be diminished. The synthesis of ammonia is exothermic; hence to obtain a high percentage conversion to ammonia, the temperature should be low. In practice, temperatures of about 550°C are used. The use of a temperature much lower than this is not practicable because the rate of combination of nitrogen and hydrogen, and thus the rate of attainment of equilibrium, would be too slow. As it is, catalysts have to be used to increase the rate of reaction. The distinction between the effect of temperature on the rate of reaction and its effect on the composition of the equilibrium mixture should be clearly borne in mind. The former follows the empirical relation

$$\log k = A - B/T \quad \text{when } A \text{ and } B \text{ are constants.}$$

The effect of temperature on the equilibrium position is subject, as stated above, to le Chatelier's principle.

Consider now the effect of changes in pressure on the equilibrium position. In the synthesis of ammonia, the combination of hydrogen and nitrogen to form ammonia reduces the number of molecules present, hence, in accordance with Avogadro's principle, there will be a decrease in pressure. The use of higher pressures will accordingly increase the proportion of ammonia in the equilibrium mixture. It will be recalled that pressures of 200 atmospheres or more are used in the Haber process.

In 1864, through the study of a reversible chemical reaction, Guldberg and Waage discovered the Equilibrium law. Berthollet had observed the formation, on the shores of hot salt lakes in Egypt, of crystals of sodium carbonate which had been formed as a result of the reaction

$$2 \text{ NaCl} + \text{CaCO}_3 \rightarrow \text{Na}_2\text{CO}_3 + \text{CaCl}_2,$$

a reaction which occurs in the reverse direction under more usual conditions.

Consider the reversible change

$$A + B \rightleftharpoons C + D.$$

According to the equilibrium law

$$\frac{[C][D]}{[A][B]} = K$$

where $[A]$ represents the 'active mass' or 'activity' (approximately equal to the partial pressure for gases and to the concentration for solutes) of A, etc., and K is a constant, the equilibrium constant. This relation enables us to predict the effect on the equilibrium position of a change in the concentration of a reactant or, in the case of gas reactions, of a change of pressure. It should be borne in mind that although changes in concentration and pressure can alter the equilibrium position, the value of K remains the same (provided the temperature is constant). Changes in temperature change the value of K, the equilibrium constant, in the following manner.

The relation $\Delta G^o = - RT \log_e K$ between K and the standard free energy of a reaction occurring in a cell, was derived on p. 236. The value of ΔG^o for an equilibrium between gases can be measured in other ways. From the relation $\Delta G^o = \Delta H^o - T\Delta S^o$, referred to on p. 208, we can relate K, the equilibrium constant, to the heat of the reaction.

If K_p represents the equilibrium constant for a gas reaction (in which the 'active masses' are measured by the partial pressures of the gases) then it follows from the above two relationships that

$$\Delta G^o = - RT \log_e K_p = \Delta H^o - T\Delta S$$

$$\therefore \log_e K_p = - \frac{\Delta H^o}{RT} + \text{a constant}$$

since, within limits, both ΔH^o and ΔS^o are largely independen: of temperature (see p. 257).

Consider again the synthesis of ammonia as an example:

$$\text{N}_2 + 3\text{H}_2 \rightleftharpoons 2\text{NH}_3, \quad K_p = \frac{[\text{NH}_3]^2}{[\text{N}_2][\text{H}_2]^3}.$$

Suppose that the nitrogen and hydrogen in the initial mixture are present in the molecular proportion of $1:3$, and that at equilibrium a fraction, x, of the nitrogen is present as ammonia. The relative number of molecules of the gases present in the equilibrium mixture will then be

$$N_2 \;+\; 3H_2 \rightleftharpoons 2NH_3 \quad \text{Total}$$
$$1-x \quad 3(1-x) \quad 2x \quad (4-2x)$$

Since, according to Avogadro's principle, the number of molecules is proportional to the partial pressure of the gas. If the total pressure is P, the partial pressures will be

$$\text{Nitrogen } \frac{1-x}{4-2x}\,P, \quad \text{Hydrogen } \frac{3(1-x)}{4-2x}\,P, \quad \text{Ammonia } \frac{2x}{4-2x}\,P.$$

Hence

$$K_p = \frac{\left(\dfrac{2x}{4-2x}\right)^2 P^2}{\dfrac{1-x}{4-2x}\,P\left(\dfrac{3(1-x)}{4-2x}\right)^3 P^3}$$

$$\therefore\; K_p = \frac{4x^2(4-2x)^2}{27(1-x)^4\,P^2}.$$

This equation relates the total pressure to the percentage conversion to ammonia. For small percentage conversions, the equation simplifies to $K_p = \dfrac{64\,x^2}{27\,P^2}$, so that x is proportional to P, i.e. for low yields, a twofold increase in pressure will approximately double the yield. This is demonstrated by the figures shown in the following table:

Temperature (°C)	Percentage conversion to ammonia		
	1 atm	100 atm	200 atm
550	0·0769	6·7	11·9
650	0·0321	3·02	5·71
750	0·0159	1·54	2·99
850	0·0089	0·87	1·68
950	0·0055	0·54	1·07

This subject was briefly discussed in Chapter 2 (p. 42). Thermal dissociation is a reversible reaction and at any given temperature a state of equilibrium exists between the undissociated molecules and the products of dissociation. As shown previously, the degree of dissociation can be obtained by means of density measurements:

$$\alpha = \frac{\rho_t - \rho_0}{\rho_0}.$$

The equilibrium constant for the dissociation can then be calculated as follows. Consider a dissociation in which one molecule dissociates into two others:

$$AB \rightleftharpoons A + B.$$

The partial pressure of the undissociated substance is $\dfrac{1 - \alpha}{1 + \alpha} P$, and the

partial pressures of the products of dissociation are $\dfrac{\alpha}{1 + \alpha} P$. Hence, the

equilibrium constant, K, is given by

$$K = \frac{\dfrac{\alpha}{1 + \alpha} P \, \dfrac{\alpha}{1 + \alpha} P}{\dfrac{1 - \alpha}{1 + \alpha} P} = \frac{\alpha^2}{1 - \alpha^2} P.$$

The values of the degree of dissociation at several different pressures are calculated from measurements of density and the results can be used to test the above equation. By repeating the measurements at different temperatures, the variation of K with T can be investigated.

In Expt. 2-6, values of α for the equilibrium $2NO \rightleftharpoons N_2O_4$ were measured at various temperatures. These results could be used to calculate corresponding values of K_p and, by plotting $\log K$ against $1/T$, to test the relation $\log_e K_p = -\dfrac{\Delta H^\circ}{RT} +$ a constant, since ΔH° is almost independent of temperature.

It follows that

$$\frac{d (\log_e K)}{dT} = \frac{\Delta H^\circ}{RT^2}$$

or

$$\log_e \frac{K_1}{K_2} = -\frac{\Delta H^\circ}{R} \left(\frac{1}{T_1} - \frac{1}{T_2} \right)$$

for two temperatures T_1 and T_2.

An approximate value of ΔH° could thus be obtained.

(b) Equilibrium in solutions of electrolytes

THE STRENGTHS OF ACIDS

Expt. 12b-1

(i) Compare the relative strengths of some acids by comparing the rates of evolution of hydrogen when the acids attack zinc. Cut some zinc sheet into squares of equal area (about 3 cm^2). Clean them with emery paper and wash them with distilled water. Fill a gas burette with 2M-hydrochloric acid and invert it over the same acid in a dish. Introduce one of the pieces of zinc into the lower end of the tube and measure the rate of evolution of hydrogen. Repeat the measurement using in turn, M solutions of M-sulphuric, $\frac{3}{2}$M-phosphoric and 2M-acetic acids, all of which contain the same number of moles of hydrogen per litre.

The magnitude of the rates will depend upon the purity of the zinc. Greater rates can be obtained by adding small, equal quantities of dilute copper sulphate solution to each acid. The experiment is only intended to give a *qualitative* comparison of the strengths of the acid. The results shown in Fig. 113 illustrate the sort of comparison that is obtained.

(ii) Make an approximate comparison of the electrical conductances of the same solutions by means of the apparatus described in Expt. 11c–2.

Expt. 12b-2

The rate of production of iodine by the reaction between acidified potassium iodide and potassium bromate,

$$HBrO_3 + 6HI \longrightarrow 3I_2 + HBr + 3H_2O$$

depends upon the hydrogen-ion concentration. Thus the rate of formation of the blue iodine-starch complex can be used to demonstrate the relative strengths of acids.

Dissolve about 2 g each of potassium bromate and potassium iodide in about 1 litre of distilled water, add a little starch solution and put 100 cm^3 in each of five beakers. Add 10 cm^3 of M/5 solutions of the following acids: hydrochloric, sulphuric, oxalic, chloroacetic, and acetic. Compare the times taken for the blue colour to appear in each case. In order to compare the strengths of the two strongest acids, sulphuric and hydrochloric, it is desirable to repeat the experiment using 10 cm^3 of M/100-acids.

Specimen results

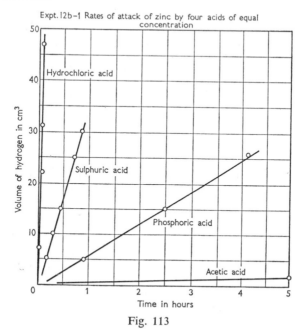

Expt. l2b-1 Rates of attack of zinc by four acids of equal concentration

Fig. 113

Ostwald's dilution law

We have already seen how the ionic dissociation theory of Arrhenious attributes the difference in strength of, say, chloroacetic acid and acetic acid of the same molarities to different degrees of ionization. We have also seen how the increased conductance on dilution is interpreted as an increased degree of dissociation. By applying the equilibrium law to the ionic dissociation

$$HA + H_2O \rightleftharpoons H_3O^+ + A^-,$$

a relation between the degree of ionization and the dilution of a given electrolyte is obtained.

If the solution contains 1 mole in v litres, and if α is the degree of dissociation, then the concentration of the hydrogen ions, acid anions, and the un-ionized acid molecules are respectively α/v, α/v and $(1-\alpha)/v$ moles per litre. (The active mass of the water remains almost constant.) Substituting these values in the equilibrium law equation, we have

$$K_a = \frac{\alpha/v \cdot \alpha/v}{(1-\alpha)/v} = \frac{\alpha^2}{v(1-\alpha)}$$

where K_a is the equilibrium constant, and is known as the ionization constant of the acid. The relation is known as Ostwald's dilution law, and those acids and bases which follow it are known as 'weak'. It may be tested by inserting values of α obtained from conductance measurements.

Let α be the fraction of molecules ionized in a given solution of dilution v (i.e. a solution containing 1 mole in v litres). The theory assumes that the conductance is proportional to the number of ions present, hence $\Lambda_v = c\alpha$ (where c is a constant) and $\Lambda_0 = c \times 1$. Therefore $\alpha = \Lambda_v/\Lambda_0$.

In order to test Ostwald's dilution law and to determine the ionization constant of the acid and the conductance at infinite dilution, the results of conductance measurements are best treated as follows:

Substituting Λ_v/Λ_0 for α in the equation $K_a = \dfrac{\alpha^2}{v(1 - \alpha)}$, we have

$$K_a = \frac{\Lambda_v}{v} \frac{\Lambda_v}{\Lambda_0} \frac{1}{(\Lambda_0 - \Lambda_v)} \quad \text{or} \quad \left(\frac{\Lambda_0^2}{\Lambda_v} - \Lambda_0\right)K_a = \Lambda_v/v.$$

So by plotting values of $1/\Lambda_v$ against values of Λ_v/v, a straight line will be obtained if the law is obeyed. The intercepts give the values of $1/\Lambda_0$ and $\Lambda_0 K_a$, whence K_a and Λ_0 can be calculated. The measured values of Λ_v for various values of v can then be used to calculate the corresponding values of α. Some typical values are shown in the table given on p. 261.

The differences in strength of the common acids is well brought out in the tables by the range in values of K and of α for 0·5M solutions. The table also shows how closely a weak acid like acetic acid follows Ostwald's dilution law and gives confirmation that the ionic dissociation theory of Arrhenius is a fair picture of the behaviour of such weak electrolytes. However, the table shows that the figures for potassium chloride do not give a constant value for K_a, and it is found that all electrolytes of high conductance give results that diverge considerably from the law in all but extremely dilute solutions. In these cases it is evident that the ionic dissociation theory does not fit the facts. Electrolytes which do not follow Ostwald's dilution law are known as *strong electrolytes*, and their behaviour is interpreted in terms of Debye and Hückel's Theory of Strong Electrolytes.

DISSOCIATION CONSTANT, K $\times 10^5$, AT 25°C		DEGREES OF DISSOCIATION α, AT 25°C		
			0·5M	0·001M
Nitrous	46	Hydrochloric	0·862	0·993
Formic	21·4	Nitric	0·862	0·997
Acetic	1·8	Trichloroacetic	0·760	0·990
Monochloroacetic	155	Monochloroacetic	0·054	0·692
Dichloroacetic	5,140	Formic	0·020	0·368
Trichloroacetic	121,000	Acetic	0·006	0·126
Benzoic	6·2	Hydrocyanic	0·00005	0·0011
Hydrocyanic	0·00007			
Phenol	0·00001			

CONDUCTANCES OF ACETIC ACID AT 25°C

v (litres)	Λ_v	α	$K_a \times 10^5$
13·57	6·068	0·0157	1·845
54·28	12·09	0·0312	1·849
108·56	16·98	0·0434	1·849
434·2	33·22	0·0857	1·849
1737·0	63·60	0·1641	1·854
3474·0	86·71	0·2236	1·855

CONDUCTANCES FOR POTASSIUM CHLORIDE AT 18°C

v (litres)	Λ_v	α	$\dfrac{\alpha^2}{(1-\alpha)v}$	$\sqrt{v}(\Lambda_v - \Lambda_0)$
1	78	0·76	2·34	32
10	112	0·86	0·54	51
50	120	0·92	0·22	71
100	122	0·94	0·15	76
500	126	0·97	0·07	83
1000	127	0·98	0·05	85
5000	129	0·99	0·02	85
∞	130	1·00	—	—

The theory of complete ionization

Many substances exist in the solid state as an aggregate of ions, not of neutral atoms. For example, in a crystal of common salt, each sodium ion is surrounded by six chlorine ions, and each chlorine ion is surrounded by six sodium ions. When the salt dissolves in water, the crystal structure breaks up and the ions move freely about in the solution. But some relic of the previous arrangement is retained, for, owing to the electrical attraction of opposite charges, each ion will tend to have an excess of the oppositely charged ions in its neighbourhood. As the ions move under an applied potential difference, the stream of anions moving towards the anode will be hindered by the stream of cations moving in the opposite direction. The extent of this opposition will be less in more dilute solutions, and the ionic velocities, and hence the conductances, will increase with dilution. It is in this way that the theory of complete ionization attempts to account for the behaviour of strong electrolytes. In 1927, an expression relating the conductance of a solution (k) with the dilution was obtained theoretically by Onsanger: $\Lambda_v - \Lambda_0 = k/\sqrt{v}$. This is identical with the empirical equation suggested by Kohlrausch in 1907. The figures in the last column of the table show to what extent the equation fits the measurements for potassium chloride.

The distinction between *weak* and *strong* electrolytes is blurred. On the whole, weak electrolytes are covalent compounds which are capable of ionizing to some extent. On the whole, strong electrolytes, with their high conductance, are ionic compounds. But not all fully ionized compounds form strong electrolytes in solution, for *ion pairs* may form and reduce the conductance. In other words, an electrolyte may be fully ionized but there may still not be complete ionic dissociation. For example, $La^{3+}Fe(CN)_6^{3-}$ is fully ionized but only slightly dissociated, whereas Na^+Cl^- is almost fully ionized and also almost fully dissociated in solution.

Expt. 12b-3   Conductance of a weak acid

A weak acid behaves in accordance with the electrolytic dissociation theory, and values of Λ_0 and K_a can be obtained by measuring the conductances of solutions of succinic or monochloroacetic acids. Prepare an M/10 solution of the pure acid in freshly distilled water and determine its exact concentration by titration with standard alkali, using phenolphthalein as indicator. Using the apparatus described in Expt. 11c-3, measure the conductance of the M/10 solution and those of solutions prepared from it by successive dilution. M/20, M/40, . . ., M/320 are suitable concentrations. By plotting Λ_v/v against $\dfrac{1}{\lambda_v}$, obtain values for Λ_0 and K_a (see p. 260). For monochloroacetic acid at 25 °C, $\Lambda_0 = 392$ and K_a is about $1 \cdot 5 \times 10^{-3}$. For succinic acid, $\Lambda_0 = 381$, and K_a is about $3 \cdot 0 \times 10^{-5}$. (This applies to the dissociation of one hydrogen ion; the second dissociation is small enough in comparison to be neglected.)

Specimen result

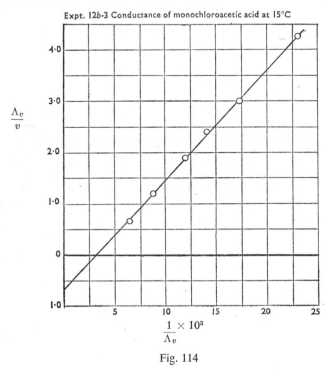

Expt. 12b-3 Conductance of monochloroacetic acid at 15°C

Fig. 114

The results for monochloroacetic acid shown in Fig. 114 give a value of 357 for Λ_0 at 15°C and $1 \cdot 7 \times 10^{-3}$ for K_a.

THE COMMON ION EFFECT

Expt. 12b-4

Place about 20 g of calcium carbonate (precipitated chalk) in each of two 500 cm^3 measuring cylinders. Prepare (a) 100 cm^3 of 2M-acetic acid, and (b) 100 cm^3 of 2M-acetic acid saturated with sodium acetate. Pour these solutions into the cylinders and note the difference in the rates of evolution of carbon dioxide.

Expt. 12b-5

Dissolve small quantities of potassium nitrite and potassium iodide in about 100 cm³ of distilled water and divide into two parts. To one add dilute acetic acid, and to the other add dilute acetic acid which has been saturated with sodium acetate. Observe the difference in the rates of production of iodine in the two solutions.

Expt. 12b-6

Use the apparatus described in Expt. 11c-2, but include a milliammeter in the circuit. Place dilute acetic acid, coloured with methyl orange, in the cell until the lamp just glows. A current of about ¼ amp is suitable. Dissolve enough sodium acetate to cover a penny in about 50 cm³ of distilled water and add some urea to increase the density. Place this solution in a tap funnel and run it slowly into the cell so that it forms a small separate layer beneath the acetic acid. Switch on the current and read the ammeter. Mix the solutions by stirring and note that the glow of the lamp becomes slightly less bright and that the ammeter reading falls.

The observations made in Expts. 12b-4 and 5 suggest that the addition of sodium acetate to acetic acid reduces the 'acidity' of the latter. The interpretation of this effect by the theory of ionic dissociation is as follows. In a solution of acetic acid, the following equilibrium exists between the ions and the molecules:

$$H A + H_2O \rightleftharpoons H_3O^+ + A^-.$$

The effect of adding sodium acetate is to increase the concentration of acetate ions and thus to shift the equilibrium towards the left. The concentration of hydrogen ions is therefore reduced. This is manifested in Expt. 12b-4 as a decrease in the rate of decomposition of the chalk, and in Expt. 12b-5 as a decrease in the rate of the reaction

$$2HNO_2 + 2HI \rightarrow I_2 + 2NO + 2H_2O.$$

The shift in the equilibrium decreases the total number of ions present and Expt. 12b-6 is an attempt to demonstrate this:

$$HA + H_2O \rightleftharpoons H_3O^+ + A^-$$
$$Na^+A^- \rightarrow Na^+ + A^-.$$

INDICATORS

Expt. 12b-7

By diluting M/10-hydrochloric acid, prepare solutions of the following concentrations and place them in boiling-tubes or beakers: 0·01M, 0·00M, 0·0001M and 0·00001M. Then dilute some M/10-caustic soda in a similar manner. To each of the ten solutions so obtained add a few drops of methyl orange, and to a duplicate set add a few drops of phenolphthalein. Note the hydrogen-ion concentration at which the indicators change colour.

Expt. 12b-8

Prepare a dilute solution of methyl violet. Add a few drops to each of eight boiling-tubes containing hydrochloric acid of the following molarities: 3M, 2M, M, 0·5M, 0·1M, 0·05M, 0·01M and 0·001M. Note the range of colours produced.

Expt. 12b-9

Place 20 cm³ of 3M-hydrochloric acid in a conical flask and add a few drops of methyl violet as indicator. Run in M-caustic soda from a burette and compare the colour changes with those in Expt. 12c-8.

Expt. 12b-10

Add a few drops of methyl violet to boiling-tubes containing solutions of 0·1M-hydrochloric, 0·1M-acetic and 0·05M-oxalic acids, and compare the colours produced with those formed in Expt. 12c-8. Estimate the hydrogen-ion concentration in these solutions.

The first two experiments above show that a given indicator changes colour over a certain hydrogen-ion concentration range, and that this concentration range may be different for different indicators. The methyl violet has several colour changes, each occurring at a different hydrogen-ion concentration. It is also evident that the change from one colour to the other is only complete over a fairly wide range of hydrogen-ion concentration, i.e. approximately 100-fold. Expt. 12b-10 enables us to measure the hydrogen-ion concentrations in comparable solutions of acetic and oxalic acids by comparing the colours of the indicator with those in Expt. 12b-8.

Ostwald's theory of indicators

The theory assumes that indicators are weak acids or bases and that the colours of the ion and of the undissociated molecule are different, $HA \rightleftharpoons H^+ + A^-$. In acid solution, the ionic dissociation will be suppressed, the indicator will be largely in the un-ionized form and of the corresponding colour. In alkaline solution, the equilibrium will be displaced by the removal of hydrogen ions to form water, so that the indicator is largely ionized. The colour of the indicator is intermediate between the two extremes when its molecules and ions are present in equal concentrations.

If K_i is the dissociation constant of the indicator, we have:

$$K_i = \frac{[H^+][A^-]}{[HA]}.$$

At the colour change point, $[A^-] = [HA]$, and so $K_i = [H^+]$, i.e. any particular indicator changes at a characteristic hydrogen-ion concentration, and this is numerically equal to the dissociation constant of the indicator.

It is usual to assume that the change to one colour is complete if the concentration of the species of that colour is ten times that of the other, i.e. if $[A^-] = 10[HA]$. Under these conditions, the hydrogen-ion concentration will be 10 times that at the change point, so the complete change from one colour to the other will take place over a hydrogen-ion concentration change of 100-fold. This was demonstrated in Expt. 12b-7. The approximate values of K_i for a few indicators are shown below:

> methyl orange 10^{-4}
> bromothymol blue 10^{-7}
> phenolphthalein 10^{-9}.

These values are numerically in the centre of the range of hydrogen-ion concentration over which the indicator changes colour.

Conductance measurements show that even in the purest water there is a small concentration of ions. These are formed by the ionization of the water:

$$H_2O \rightleftharpoons H^+ + OH^-.$$

The hydrogen ions are almost entirely in the form of hydroxonium ions but are often represented by the symbol H^+ for convenience.

$$2 H_2O \rightleftharpoons H_3O^+ + OH^-.$$

At room temperature the ionic product of water, $[H^+][OH^-]$, is about 10^{-14}. In pure water, the hydrogen-ion and hydroxyl-ion concentrations are equal, hence $[H^+] = 10^{-7}$ moles per litre. When $[H^+] = [OH^-]$, a solution is neither acidic nor alkaline and is said to be 'neutral'. When a solute is dissolved in the water, the equilibrium position of the ionic dissociation may be affected. For example, the addition of an acid increases the hydrogen-ion concentration and the hydroxyl-ion concentration is correspondingly reduced. The addition of the salt of a strong acid and a weak base, or of a strong base and a weak acid, also changes the hydrogen-ion concentration. For example, sodium acetate solutions are alkaline because the acetate ions combine with hydrogen ions to form acetic acid molecules, thus resulting in an increase in the hydroxyl-ion concentration. The acidity of ammonium chloride solutions can be interpreted in a similar manner. A concentrated solution of ammonium chloride evolves hydrogen when heated with a zinc-copper couple, and liberates carbon dioxide from calcium carbonate.

A useful nomenclature for hydrogen-ion concentrations makes use of a logarithmic scale and is called the 'pH scale'. Thus, a solution in which the hydrogen-ion concentration is 10^{-x} gram-ions per litre, is said to have a pH of x. The pH of a 0·01M solution of hydrochloric acid is about 2. The pH of neutral water is about 7 and the pH of a 0·01M-caustic soda solution is 12. Methyl orange changes colour in a solution of about pH 4·6, and phenolphthalein in a solution of about pH 9.

Expt. 12b-11 Measurement of pH by means of indicators

The pH of tap water or of soil-drainage water may be conveniently measured in this way. First determine by trial the approximate pH with the indicators listed in the table. Then select the indicator within whose pH range the acidity of the test solution lies. The pH is then measured more exactly by comparing the colour of the indicator when present in the test solution with a set of colour standards. These consist of pairs of tubes, one containing acid and the other alkali, to which a total of, say, 10 drops of indicator have been added. The ratio of the number of drops of indicator in the two tubes varies from pair to pair over a range of 1 : 9 to 9 : 1. If the pairs of tubes are then viewed end-on, their colours range from that of the indicator in acid solution through intermediate shades to the colour exhibited in alkaline solution. The colours formed by the combination of the two extreme colours in various ratios correspond to the colours which the indicator would assume in solutions of

certain values of pH. These values for seven indicators and nine
drop ratios are shown in the table below.

Select twenty $\frac{1}{2}$ in. test-tubes of uniform internal diameter. Drill
three pairs of vertical holes in a wooden block to hold the tubes,
and drill three horizontal holes, each passing through a pair of the
vertical holes, for viewing the tubes. Make stock solutions of the
indicators in the prescribed manner. For use, dilute these stock
solutions tenfold. Place 5 cm^3 of distilled water in all the tubes but
one, and in this put 5 cm^3 of the solution under test. Place the
requisite number of drops of indicator solution in nine pairs of
tubes to give the drop ratios listed in the table, and place ten drops
of indicator in the test solution. Turn the indicator to its 'acid'
colour in one series of nine tubes by adding one drop of M/20-
hydrochloric acid to each and make the second set alkaline by
adding a drop of M/20-NaOH to each. Match the colour transmitted by the test solution and the tube of distilled water with that
transmitted by one of the pairs of standards. Read off the pH from
the table. Use the method to determine the pH of decimolar solutions of salts such as Na_2HPO_4, $MgSO_4$, NH_4Cl, etc.

Drop ratio	Bromphenol blue	Methyl red	Bromcresol purple	Bromthymol blue	Phenol red	Cresol red	Thymol blue
1 : 9	3·1	4·1	5·3	6·2	6·8	7·2	7·9
2 : 8	3·5	4·4	5·7	6·5	7·1	7·5	8·2
3 : 7	3·7	4·6	5·9	6·7	7·3	7·7	8·4
4 : 6	3·9	4·8	6·1	6·9	7·5	7·9	8·6
5 : 5	4·1	5·0	6·3	7·1	7·7	8·1	8·8
6 : 4	4·3	5·2	6·5	7·3	7·9	8·3	9·0
7 : 3	4·5	5·4	6·7	7·5	8·1	8·5	9·2
8 : 2	4·7	5·6	6·9	7·7	8·3	8·7	9·4
9 : 1	5·0	6·0	7·2	8·0	8·7	9·1	9·8

Expt. 12b-12   Use of indicators in titrations

Test dilute solutions of (a) sodium carbonate, (b) potassium cyanide,
(c) ammonium chloride, (d) iron(II) sulphate, (e) copper sulphate
(all in distilled water) with litmus solution. It will be found that
(a) and (b) are alkaline to litmus and the others are acid.

In an acid-alkali titration, an indicator is used to determine, not the
neutral point, but the *equivalence point*. When an equivalent of a strong

acid has been added to an equivalent of a strong base, the resulting solution is neutral, but if an equivalent of a weak acid is used, the resulting solution is not neutral but alkaline. Hence to determine the equivalence point, an indicator which changes colour on the alkaline side of neutrality must be used. Again, to determine the equivalence point when titrating a weak alkali against a strong acid, an indicator

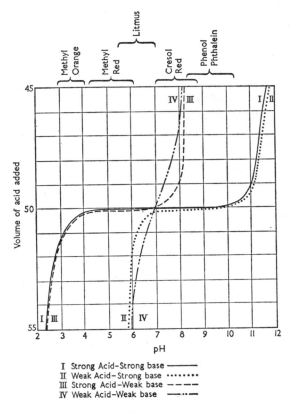

I Strong Acid–Strong base ————
II Weak Acid–Strong base ••••••••
III Strong Acid–Weak base — — — —
IV Weak Acid–Weak base —••—••—

Fig. 115

that changes on the acid side of neutrality is required. The changes in pH that occur in the course of acid-alkali titrations are shown in Fig. 115. From such diagrams, and a knowledge of the pH range over which each indicator changes, suitable indicators for specific titrations can be selected.

These phenomena are interpreted in terms of the ionic theory as follows. The reaction between an acid and a base to form a salt and water may be a reversible reaction, the reverse change being called 'hydrolysis'. Salts of strong bases and strong acids are not appreciably hydrolysed in water, the reaction $HA + BOH \longrightarrow BA + H_2O$ is irreversible. Salts of a weak base with a strong acid (such as ammonium sulphate, aniline hydrochloride, etc.) or of a strong base with a weak acid (such as sodium sulphide, sodium acetate, etc.) are partially hydrolysed, resulting in non-neutral solutions. For such salts, the reaction $HA + BOH \rightleftharpoons BA + H_2O$ is reversible.

Weak base $B^+A^- + H_2O \longrightarrow BOH + H^+ + A^-$ acidic,

Weak acid $B^+A^- + H_2O \longrightarrow B^+ + OH^- + HA$ alkaline.

The extent of the hydrolysis varies with the strength of the weak acid or base, and with the concentration of the solution (see Expt. 12b-13(ii)).

It is interesting to recall reactions in which hydrolysis takes place to completion, in which the reaction

$$BA + H_2O \rightleftharpoons HA + BOH$$

is irreversible, e.g. $PCl_3 + 3H_2O \longrightarrow 3HCl + P(OH)_3$.

Buffer solutions

Solutions of constant pH are often required in chemical and biological work. If a very dilute solution of caustic soda or hydrochloric acid, of say pH 8 or 6 respectively, is kept in a stoppered glass bottle, the pH does not remain constant for long. The slow hydrolysis of the glass surface liberates alkali; carbon dioxide may enter from the air. Expt. 12b-13 (vi) below shows how solutions may be made whose pH remains sensibly constant in spite of considerable additions of hydrogen or hydroxyl ions.

The principle on which these so-called *buffer solutions* is based is that if a solution contains a weak acid and one of its salts, the salt provides a reservoir of anions which combine with any added hydrogen ions, and 'lock them up' as undissociated acid molecules. The pH of the solution thus remains sensibly unchanged:

$$H^+ + A^- \rightleftharpoons HA.$$

On adding acid:

$$Na^+ + A^- + H^+ + Cl^- \longrightarrow Na^+ + Cl^- + HA.$$

The buffered solution also resists an increase in hydroxyl-ion concentration by virtue of the reaction:

$$HA + OH^- \longrightarrow A^- + H_2O.$$

Expt. 12b-13 Experiments with a universal indicator

Use B.D.H. Universal Indicator (pH 3–11). This indicator assumes the following colours: pH 3 red, pH 4 deep red, pH 5 orange, 6 orange-yellow, 7 green-yellow, 8 green, 9 blue, 10 violet, 11 deep violet. Use two drops to every 10 cm³ of solution tested.

(i) *Dilution of a strong acid.* Dilute some M/10-hydrochloric acid to M/100 with distilled water, and repeat the tenfold dilution four times, to obtain M/1000, M/10,000, M/100,000 and M/1,000,000 solutions. Test these with the indicator, which should show the pH values 3, 4, 5, 6.

(ii) pH *of salt solutions.*

Electrolyte	g-mols/litre	pH	Colour of indicator
NH_4Cl	2·0	4·0	Red
NH_4Cl	0·1	5·0	Orange
$MgSO_4$	0·1	6·0	Yellow
NaCl	1·0	7·0	Green-yellow
Na_2HPO_4	0·1	9·0	Blue
Na_2CO_3	0·1	11·6	Purple

(iii) *The common ion effect.* To 50 cm³ of M/10-ammonium hydroxide solution, add 1 cm³ of indicator. Then add solid ammonium carbonate, about 2 g at a time, and note the changes in pH produced.

(iv) *Hydrolysis of an ester.* Into a 1-litre flask put 500 cm³ of water, 5 cm³ of the indicator and 10 cm³ of saturated barium hydroxide solution. Then add 5 cm³ of a suitable ester (methyl formate or freshly redistilled methyl acetate) and mix well. Owing to the hydrolysis of the ester the colour changes from purple through indigo to blue, green and yellow, and after some time to orange and red.

(v) *Changes of* pH *at electrodes during electrolysis.* Fill a U-tube, fitted with platinum electrodes, with sodium sulphate solution, and add enough indicator to give a deep green colour. Use two 2 V accumulators and observe the colour changes around the electrodes as electrolysis proceeds.

(vi) *Buffer solutions.* Buffer solutions can be made by mixing appropriate quantities of the solutions listed below:

BUFFER SOLUTIONS[1]

pH	soln A cm³	soln B cm³	pH	soln C cm³	soln D cm³
*10·0	25	21·8	8·0	1·4	50
9·6	25	18·0	*7·6	3·4	50
9·2	25	12·8	7·2	5·2	25
*8·8	25	8·6	6·8	7·4	25
8·4	25	4·0	6·4	11·1	25
			*6·0	14·6	25
			5·6	18·1	25
			5·2	21·6	25
			*4·8	25	24·3
			4·4	25	19·8
			4·0	25	15·7
			*3·6	25	11·9
			3·2	25	8·2
			2·8	25	4·7
			*2·4	25	2·2

Solution A 0·1M-boric acid, prepared by dissolving 3·09 g of AR boric acid and 3·73 g potassium chloride in water and making up to 500 cm³.

Solution B 0·1M-sodium hydroxide.

Solution C 0·1M-citric acid, prepared by dissolving 9·60 g of AR citric acid in water and making up to 500 cm³.

Solution D 0·2M-disodium hydrogen phosphate, prepared by dissolving 17·82 g of pure crystals, $Na_2HPO_4, 2H_2O$, in water and making up to 500 cm³.

Place 10 cm³ of each of the solutions marked with an asterisk in separate boiling-tubes, and put 10 cm³ of distilled water in an eighth tube. Add five drops of indicator to each tube, and pour about a third of the liquid out of each boiling-tube into test-tubes to serve as reference colours. Then add one drop of M/10-hydrochloric acid to each boiling-tube. Note that the acidity produced in the distilled water is considerable, whereas the pH change in the buffer solutions is hardly observable.

Repeat the experiment, adding one drop of M/10-caustic soda instead of the drop of hydrochloric acid.

---

[1] From *Advanced Level Practical Chemistry* by J. W. Davis (John Murray, 1963).

Expt. 12b-14   Investigation of the equilibrium between the two forms of an indicator—phenolphthalein

Prepare buffer solutions of the following values of pH by adding the appropriate volumes of solution B (above) to 25 cm³ of solution A.

pH	$[H^+] \times 10^{10}$	cm³ of soln B
9·0	10	10·7
9·23	6	13·1
9·4	4	15·5
9·5	3·2	16·9
9·7	2	18·9
10·0	1	21·8

Place 25 cm³ of each buffer solution in a conical flask or beaker and add exactly one drop of the phenolphthalein solution to each. The drops must all be of the same size. Using flat-bottomed test-tubes, compare the intensity of colours in pairs of the solutions.

Specimen result

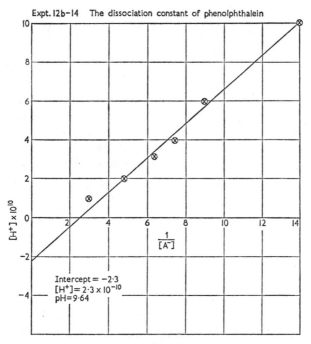

Expt. 12b-14   The dissociation constant of phenolphthalein

Intercept = −2·3
$[H^+] = 2·3 \times 10^{-10}$
pH = 9·64

Fig. 116

EBC—T

Do this by adjusting the depths of the solutions in the test-tubes and looking vertically down through the solutions on to a white tile. These comparative figures are a measure of the relative concentrations of the coloured ion in the solution

$$HA \rightleftharpoons H^+ + A^-$$

colourless   coloured

$$K_i = \frac{[H^+][A^-]}{[HA]}.$$

The total concentration of indicator present, S, is $[A^-] + [HA]$.

So, $K_i(S - [A^-]) = [H^+][A^-]$

and $K_i S \dfrac{1}{[A^-]} - K_i = [H^+].$

By plotting $[H^+]$ against $1/[A^-]$, using the depths of the solutions in the test-tubes as a measure of the latter, a straight line should be obtained. The intercept on the $[H^+]$ axis should equal $-K_i$ (about 9·6).

(c) Heterogeneous equilibria

Expt. 12c-1

Pass a stream of carbon dioxide over a few crystalline, transparent flakes of calcite heated strongly in a combustion tube. Note that they remain unchanged. Allow to cool, remove a few crystals and add a drop of water to show that no quicklime has formed. Replace the carbon dioxide by a slow current of air. The crystals will be observed to change in appearance. After about 10 minutes allow to cool, remove the crystals and add a drop of water:

$$CaCO_3 \rightleftharpoons CaO + CO_2.$$

The behaviour of reacting substances that are not all in the same phase calls for separate treatment. The laws of equilibrium that apply in one phase do not necessarily do so when the equilibrium is set up across a phase boundary.

As an example consider Expt. 12c-1. It demonstrates the effect of the carbon dioxide pressure on the decomposition of calcium carbonate. It is known that the pressure of carbon dioxide in equilibrium with the $CaCO_3$ and CaO is constant at any particular temperature. At red heat the dissociation pressure of the system $CaCO_3 \rightleftharpoons CaO + CO_2$ is less than one atmosphere, and therefore calcium carbonate will not decompose at that temperature in a stream of carbon dioxide at atmospheric pressure. However, if the carbon dioxide pressure is kept low by the passage of a current of air, the calcium carbonate rapidly decomposes.

In a similar manner the formation of mercury(II) oxide from mercury and oxygen depends on the use of appropriate temperatures and pressures:

$$2HgO \rightleftharpoons 2Hg + O_2.$$

The dissociation pressures at a number of temperatures are shown below. It is clear that the dissociation pressure will exceed the partial pressure of oxygen in the air (about 150 mm of mercury) at a temperature a few degrees above 360 °C. It would therefore not be possible to make mercury oxide by heating mercury in the air above this temperature. However, mercury boils at 357°C, so that if mercury is heated to its boiling point in air, it is possible for the oxide to form. Were the dissociation pressure of mercury oxide somewhat higher than it is, Lavoisier's historic experiment on the synthesis of the oxide would not have been possible. In view of Lavoisier's deductions from this experiment about the nature of the air and of combustion, it is interesting to speculate what effect this might have had on the course of chemical discovery.

Mercury oxide can, of course, be more readily synthesized by using a higher pressure of oxygen than exists in the atmosphere.

The dissociation pressures of some other metallic oxides at various temperatures are shown below.

	Temp. (deg C)	Pressure (mm mercury)
ZnO	1227	$2 \cdot 7 \times 10^{-14}$
	3427	76
PbO	900	$3 \cdot 2 \times 10^{-18}$
	1500	$4 \cdot 8 \times 10^{-8}$
	2000	$3 \cdot 7 \times 10^{-4}$
CuO($\rightarrow$ Cu$_2$O)	1000	118
	1050	314
HgO	360	90
	420	387
	480	1581
Ag$_2$O	25	0·38
	125	152
	200	1330

The oxide will decompose at the temperature at which its dissociation pressure exceeds the partial pressure of the oxygen in the air (i.e. about one-fifth of an atmosphere or 150 mm of mercury). The figures clearly show that the oxides of the more electropositive metals are extremely stable, whereas those of the nobler metals are very unstable. Thus copper(II) oxide becomes copper(I) oxide on being heated in the air to just over 1000 °C, mercury oxide decomposes at under 400 °C, and silver oxide breaks down at about 125 °C.

Expt. 12c-2　The equilibrium between silver chloride and ammonia

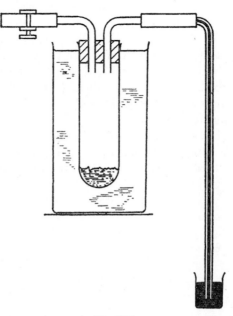

Fig. 117

Precipitate about 5 g of silver chloride, wash it with distilled water, and thoroughly dry it in a hot-air oven. Place it in a dry boiling-tube or small flask fitted with a rubber bung and connected by gas-tight joints to a simple mercury manometer and a source of dry ammonia gas (Fig. 117). Generate a stream of ammonia, well dried by fresh quicklime, and pass the gas through the apparatus, having surrounded the boiling-tube with a beaker of water at about 100 °C. Disconnect the ammonia supply and close the tube leading to it by means of a screw-clip. Replace the hot water by a beaker of water at room temperature, and note the manometer reading. When it is constant, raise the temperature of the water about 20 deg and repeat the measurement.

The system silver chloride-ammonia has the following dissociation pressures at the temperatures shown:

Temp. (deg C)	0	24	48	57
mm of mercury	30	94	240	490

The system will take some time to come to equilibrium, and this time will be longer at the lower temperatures. See how far it is possible to reproduce these figures in a simple apparatus of this kind. An agreement to within 10% has been obtained.

SOLUBILITY

Expt. 12c-3 Equilibrium between a solid and its solution

Shake a teaspoonful of sodium nitrate or other similar solute in about 50 cm³ of water in a conical flask. Note any temperature changes that occur. Does any salt remain undissolved? If so, what is the effect of warming the flask?

The dissolution of a solid in a solvent is often, but not always, an endothermic process. It is interesting to ask the question 'why do such solution processes occur?' The energy changes involved include (i) the energy needed to break the bonds holding the ions or molecules together in the solid (lattice energy), (ii) the energy needed to disperse the solute particles (largely kinetic energy), (iii) the energy of solvation of the particles, which may help to supply the energy needed for (i) and (ii). If the process of dissolution is endothermic, this indicates that energy from (iii) is insufficient for this purpose and further supplies are taken from the kinetic energy of the particles of solvent, whose temperature therefore falls.

The driving force making this endothermic process occur is clearly not the change in thermal energy content ($\Delta H°$). We have already seen (p. 208) that another factor is the change in free energy, $\Delta G°$, and that these quantities are related by the equation

$$\Delta H° = \Delta G° + T\Delta S°,$$

where $\Delta S°$ is the change in entropy. In the case of the dissolution of a solid in a liquid solvent, there is a large increase of randomness among the particles involved, i.e. an increase in entropy. So although $\Delta H°$ may be positive (i.e. the process is endothermic) $\Delta G°$ may still be negative owing to the predominating effect of the large increase in $T\Delta S°$, and therefore the process will take place (see Chap. 10).

In a saturated solution there is a heterogeneous equilibrium between solid solute and the saturated solution. This equilibrium is dynamic, i.e. is proceeding both ways and this can be demonstrated by the use of radioactive tracers (see Expt. 1d-5).

$$AB \text{ (solid)} \rightleftharpoons AB \text{ or } A^+ B^- \text{ (dissolved)}.$$

The equilibrium constant expresses the magnitude of the solubility:

$$K = \frac{[AB] \text{ (dissolved)}}{[AB] \text{ (solid) i.e. a constant}}.$$

The equilibrium constant, expressing the extent to which the solute dissolves, is related to the free energy change (the factor determining this) by the equation:

$$\Delta G° = - RT \log_e K.$$

For $T = 300° K$,

$$\Delta G \text{ (in kcals)} = - 1\cdot36 \log_{10} K.$$

It is interesting to note the quantitative consequences of this logarithmic relationship, e.g. a change of 4 kcals in the value of $\Delta G°$ changes K by a factor of 1000. In the case of the solution of an ionic solid in water, the lattice energy and the energy of hydration may both be very large; because it is the difference between these two quantities which influences $\Delta H°$ (and therefore $\Delta G°$), the solubilities of apparently rather similar compounds (e.g. $CaCl_2$ and CaF_2) may differ greatly.

Expt. 12c-4 Solubility product

Precipitate magnesium hydroxide from a few drops of a solution of magnesium sulphate in a boiling-tube by adding a little ammonium hydroxide solution. Shake the contents of the tube with a teaspoonful of solid ammonium chloride and note the effect on the precipitate of magnesium hydroxide.

The product of the concentrations of the ions of a solute in a *saturated* solution is called the *solubility product* of the solute. In a dilute solution, the product of the ionic concentrations is less than the solubility product. If the ionic concentrations are increased (e.g. by the addition of more solute, or by the addition of a substance containing a common ion), their product may be raised above the solubility product and the solute will then be precipitated. The solubility product of a substance is a measure of its solubility, and, of course, varies with temperature.

The results of Expt. 12c-4 are simply explained in terms of the above conceptions:

$$Mg^{2+} + 2OH^- \longrightarrow Mg(OH)_2;$$

$$NH_4OH \rightleftharpoons NH_4^+ + OH^-;$$

$$NH_4Cl \longrightarrow NH_4^+ + Cl^-.$$

The concentration of hydroxyl ions in the ammonium hydroxide reagent is sufficient, when multiplied by the concentration of the magnesium ions in the solution under test, to give an ionic product greater than the solubility product of magnesium hydroxide. This is therefore precipitated. But when ammonium chloride is added, the presence of extra ammonium ions depresses the ionization of the ammonium hydroxide. The hydroxyl-ion concentration is thus lowered and becomes too small for the ionic product $[Mg^{2+}][OH^-]^2$ to exceed the solubility product of magnesium hydroxide, which therefore remains in solution. Iron(III) hydroxide, which has a smaller solubility product than magnesium hydroxide, does not behave in this way; the hydroxyl-ion concentration in the mixed solution of ammonium hydroxide and chloride is sufficient to cause the precipitation of iron(III) hydroxide. This difference is made use of in the separation of these metals in analysis.

Similar explanations apply to the solubility of cadmium sulphide in strong, but not in weak, hydrochloric acid; to the precipitation of zinc sulphide in dilute acid solution if the zinc-ion concentration is great enough, and so on.

Consider the reversible reaction

$$FeS + H_2SO_4 \rightleftharpoons FeSO_4 + H_2S.$$

In terms of the ionic theory, the equilibria involved are:

$$FeS \rightleftharpoons Fe^{2+} + S^{2-} \quad \text{and} \quad H_2S \rightleftharpoons 2H^+ + S^{2-}.$$

For the latter,

$$K = \frac{[H^+]^2[S^{2-}]}{[H_2S]}.$$

If the hydrogen ion concentration is increased, the equilibrium position is moved to the left, thus decreasing the concentration of sulphur ions. A certain minimum concentration of sulphide ions is required for the solubility product of FeS to be exceeded and to precipitate it from a given solution of an iron(II) salt. It is clear, therefore, that whether iron(II) sulphide is precipitated or not depends on the acidity of the solution.

Expt. 12c-5   The solubility product of silver acetate

In this experiment saturated solutions of silver acetate are made by mixing standard solutions of silver nitrate and sodium acetate in various proportions. The silver acetate that is precipitated is removed by filtration, and the silver remaining in solution is estimated by titration with standard potassium thiocyanate.

Prepare 250 cm³ each of M/5-silver nitrate, M/5-sodium acetate, and M/10-potassium thiocyanate solutions. Mix the following quantities of the first two solutions in beakers:

M/5-silver nitrate (cm³)	50	40	30	20
M/5-sodium acetate (cm³)	30	40	50	60

Stir the solutions well until precipitation of silver acetate is complete, and then filter the solutions into four conical flasks. Titrate 20 cm³ portions of the filtrates with standard thiocyanate solution, using a little acidified iron(III) alum solution as indicator. (For details, see text books on volumetric analysis.)

For each of the four experiments calculate the following quantities in turn and thus find the solubility product of the silver acetate:
(i) The concentration of silver ions in the solution ($[Ag^+]$ gram-ions per litre).
(ii) The number of gram-ions of silver left in solution.
(iii) The number of gram-ions of silver precipitated (which equals the number of gram-ions of acetate precipitated).
(iv) The number of gram-ions of acetate left in solution.
(v) The concentration of the acetate ion in the solution.

An average value for the solubility product, $[Ag^+][Ac^-]$, that has been obtained at the temperature of the laboratory (15°C) by this method is 0·0044.

Bibliography

Allen, J. A. *Energy Changes in Chemistry* (Blackie)
Bragg, W. H. and W. L. *The Crystalline State* (G. Bell)
Browning, D. R. *Chromatography* (McGraw-Hill)
Browning, D. R. *Spectroscopy* (McGraw-Hill)
Bunn, C. J. *Crystals* (Academic Press Inc.)
Campbell, J. A. *Why Do Chemical Reactions Occur?* (Prentice-Hall Inc.)
Cartwell, E. and Fowles, G. W. A. *Valency and Molecular Structure*
    (Butterworth)
Desch, C. H. *Metallography* (Longmans)
Evans, R. C. *Introduction to Crystal Chemistry* (C.U.P.)
Firth, D. C. *Elementary Chemical Thermodynamics* (O.U.P.)
Glasstone, S. and Lewis, D. *Elements of Physical Chemistry* (Macmillan)
Hills, P. J. *Chemical Equilibria* (Edward Arnold)
King, E. L. *How Chemical Reactions Occur* (Benjamin, N.Y.)
Lowry, T. M. *Historical Introduction to Chemistry* (Macmillan)
Peacocke, T. A. H. *Atomic and Nuclear Chemistry* (Pergamon)
Read, H. H. *Elements of Mineralogy* (Allen and Unwin)
Thomson, S. J. and Webb, G. *Heterogeneous Catalysis* (Oliver and Boyd)
Trotman-Dickenson, A. F. and Parfitt, G. D. *Chemical Kinetics and Surface
    Colloid Chemistry* (Pergamon)
Van Vlack, L. H. *Elements of Materials Science* (Addison-Wesley)
Wells, M. K. *Minerals and the Microscope* (Allen and Unwin)

Conversion of calories to joules

(Correct to four significant figures)

1 cal=4·1840 J

cal	0	1	2	3	4	5	6	7	8	9
100	418·4	422·6	426·8	431·0	435·1	439·3	443·5	447·7	451·9	456·1
110	460·2	464·4	468·6	472·8	477·0	481·2	485·3	489·5	493·7	497·9
120	502·1	506·3	510·4	514·6	518·8	523·0	527·2	531·4	535·6	539·7
130	543·9	548·1	552·3	556·5	560·7	564·8	569·0	573·2	577·4	581·6
140	585·8	589·9	594·1	598·3	602·5	606·7	610·9	615·0	619·2	623·4
150	627·6	631·8	636·0	640·2	644·3	648·5	652·7	656·9	661·1	665·3
160	669·4	673·6	677·8	682·0	686·2	690·4	694·5	698·7	702·9	707·1
170	711·3	715·5	719·6	723·8	728·0	732·2	736·4	740·6	744·8	748·9
180	753·1	757·3	761·5	765·7	769·9	774·0	778·2	782·4	786·6	790·8
190	795·0	799·1	803·3	807·5	811·7	815·9	820·1	824·2	828·4	832·6
200	836·8	841·0	845·2	849·4	853·5	857·7	861·9	866·1	870·3	874·5
210	878·6	882·8	887·0	891·2	895·4	899·6	903·7	907·9	912·1	916·3
220	920·5	924·7	928·8	933·0	937·2	941·4	945·6	949·8	954·0	958·1
230	962·3	966·5	970·7	974·9	979·1	983·2	987·4	991·6	995·8	1000
240	1004	1008	1013	1017	1021	1025	1029	1033	1038	1042
250	1046	1050	1054	1059	1063	1067	1071	1075	1079	1084
260	1088	1092	1096	1100	1105	1109	1113	1117	1121	1125
270	1130	1134	1138	1142	1146	1151	1155	1159	1163	1167
280	1172	1176	1180	1184	1188	1192	1197	1201	1205	1209
290	1213	1218	1222	1226	1230	1234	1238	1243	1247	1251
300	1255	1259	1264	1268	1272	1276	1280	1284	1289	1293
310	1297	1301	1305	1310	1314	1318	1322	1326	1331	1335
320	1339	1343	1347	1351	1356	1360	1364	1368	1372	1377
330	1381	1385	1389	1393	1397	1402	1406	1410	1414	1418
340	1423	1427	1431	1435	1439	1443	1448	1452	1456	1460
350	1464	1469	1473	1477	1481	1485	1490	1494	1498	1502
360	1506	1510	1515	1519	1523	1527	1531	1536	1540	1544
370	1548	1552	1556	1561	1565	1569	1573	1577	1582	1586
380	1590	1594	1598	1602	1607	1611	1615	1619	1623	1628
390	1632	1636	1640	1644	1648	1653	1657	1661	1665	1669
400	1674	1678	1682	1686	1690	1695	1699	1703	1707	1711
410	1715	1720	1724	1728	1732	1736	1741	1745	1749	1753
420	1757	1761	1766	1770	1774	1778	1782	1787	1791	1795
430	1799	1803	1807	1812	1816	1820	1824	1828	1833	1837
440	1841	1845	1849	1854	1858	1862	1866	1870	1874	1879
450	1883	1887	1891	1895	1900	1904	1908	1912	1916	1920
460	1925	1929	1933	1937	1941	1946	1950	1954	1958	1962
470	1966	1971	1975	1979	1983	1987	1992	1996	2000	2004
480	2008	2013	2017	2021	2025	2029	2033	2038	2042	2046
490	2050	2054	2059	2063	2067	2071	2075	2079	2084	2088
cal	0	1	2	3	4	5	6	7	8	9

cal	0	1	2	3	4	5	6	7	8	9
500	2092	2096	2100	2105	2109	2113	2117	2121	2125	2130
510	2134	2138	2142	2146	2151	2155	2159	2163	2167	2171
520	2176	2180	2184	2188	2192	2197	2201	2205	2209	2213
530	2218	2222	2226	2230	2234	2238	2243	2247	2251	2255
540	2259	2264	2268	2272	2276	2280	2284	2289	2293	2297
550	2301	2305	2310	2314	2318	2322	2326	2330	2335	2339
560	2343	2347	2351	2356	2360	2364	2368	2372	2377	2381
570	2385	2389	2393	2397	2402	2406	2410	2414	2418	2423
580	2427	2431	2435	2439	2443	2448	2452	2456	2460	2464
590	2469	2473	2477	2481	2485	2489	2494	2498	2502	2506
600	2510	2515	2519	2523	2527	2531	2536	2540	2544	2548
610	2552	2556	2561	2565	2569	2573	2577	2582	2586	2590
620	2594	2598	2602	2607	2611	2615	2619	2623	2628	2632
630	2636	2640	2644	2648	2653	2657	2661	2665	2669	2674
640	2678	2682	2686	2690	2694	2699	2703	2707	2711	2715
650	2720	2724	2728	2732	2736	2741	2745	2749	2753	2757
660	2761	2766	2770	2774	2778	2782	2787	2791	2795	2799
670	2803	2807	2812	2816	2820	2824	2828	2833	2837	2841
680	2845	2849	2853	2858	2862	2866	2870	2874	2879	2883
690	2887	2891	2895	2900	2904	2908	2912	2916	2920	2925
700	2929	2933	2937	2941	2946	2950	2954	2958	2962	2966
710	2971	2975	2979	2983	2987	2992	2996	3000	3004	3008
720	3012	3017	3021	3025	3029	2033	2038	3042	3046	3050
730	3054	3059	3063	3067	3071	3075	3079	3084	3088	3092
740	3096	3100	3105	3109	3113	3117	3121	3125	3130	3134
750	3138	3142	3146	3151	3155	3159	3163	3167	3171	3176
760	3180	3184	3188	3192	3197	3201	3205	3209	3213	3217
770	3222	3226	3230	3234	3238	3243	3247	3251	3255	3259
780	3264	3268	3272	3276	3280	3284	3289	3293	3297	3301
790	3305	3310	3314	3318	3322	3326	3330	3335	3339	3343
800	3347	3351	3356	3360	3364	3368	3372	3376	3381	3385
810	3389	3393	3397	3402	3406	3410	3414	3418	3423	3427
820	3431	3435	3439	3443	3448	3452	3456	3460	3464	3469
830	3473	3477	3481	3485	3489	3494	3498	3502	3506	2510
840	3515	3519	3523	3527	3531	3535	3540	3544	3548	3552
850	3556	3561	3565	3569	3573	3577	3582	3586	3590	3594
860	3598	2602	3607	3611	3615	3619	3623	3628	3632	3636
870	3640	3644	3648	3653	3657	3661	3665	3669	3674	3678
880	3682	3686	3690	3694	3699	3703	3707	3711	3715	3720
890	3724	3728	3732	3736	3740	3745	3749	3753	3757	3761
900	3766	3770	3774	3778	3782	3787	3791	3795	3799	3803
910	3807	3812	3816	3820	3824	3828	3833	3837	3841	3845
920	3849	3853	3858	3862	3866	3870	3874	3879	3883	3887
930	3891	3895	3899	3904	3908	3912	3916	3920	3925	3929
940	3933	3937	3941	3946	3950	3954	3958	3962	3966	3971
950	3975	3979	3983	3987	3992	3996	4000	4004	4008	4012
960	4017	4021	4025	4029	4033	4038	4042	4046	4050	4054
970	4058	4063	4067	4071	4075	4079	4084	4088	4092	4096
980	4100	4105	4109	4113	4117	4121	4125	4130	4134	4138
990	4142	4146	4151	4155	4159	4163	4167	4171	4176	4180
cal	0	1	2	3	4	5	6	7	8	9

Index